Ergebnisse der Mathematik und ihrer Grenzgebiete

Band 21

Herausgegeben von P. R. Halmos · P. J. Hilton
R. Remmert · B. Szőkefalvi-Nagy

Unter Mitwirkung von L. V. Ahlfors · R. Baer
F. L. Bauer · A. Dold · J. L. Doob · S. Eilenberg
K. W. Gruenberg · M. Kneser · G. H. Müller
M. M. Postnikov · B. Segre · E. Sperner

Geschäftsführender Herausgeber: P. J. Hilton

Mahlon M. Day

Normed Linear Spaces

Third Edition

Springer-Verlag New York Heidelberg Berlin
1973

Mahlon M. Day
University of Illinois, Urbana, Illinois, U.S.A.

AMS Subject Classifications (1970): Primary 46 B xx · Secondary 46 A xx

ISBN 0-387-06148-7 Springer-Verlag New York Heidelberg Berlin
ISBN 3-540-06148-7 Springer-Verlag Berlin Heidelberg New York

ISBN 0-387-02811-0 Second edition Springer-Verlag New York Heidelberg Berlin
ISBN 3-540-02811-0 Second edition Springer-Verlag Berlin Heidelberg New York

Foreword to the First Edition

This book contains a compressed introduction to the study of normed linear spaces and to that part of the theory of linear topological spaces without which the main discussion could not well proceed.

Definitions of many terms which are required in passing can be found in the alphabetical index. Symbols which are used throughout all, or a significant part, of this book are indexed on page 132*. Each reference to the bibliography is made by means of the author's name, supplemented when necessary by a number in square brackets. The bibliography does not completely cover the available literature, even the most recent; each paper in it is the subject of a specific reference at some point in the text.

The writer takes this opportunity to express thanks to the University of Illinois, the National Science Foundation, and the University of Washington, each of which has contributed in some degree to the cultural, financial, or physical support of the writer, and to Mr. R. R. Phelps, who eradicated many of the errors with which the manuscript was infested.

Urbana, Illinois (USA), September 1957 — Mahlon M. Day

Foreword to the Third Edition

The major changes in this edition are: Corrections and additions to II, § 5, and III, § 1, on bw^* and ew^* topologies; enlargement of III, § 2, on weak compactness, V, § 1, on extreme and exposed points, V, § 2, on fixed points, and VII, § 2, on rotundity and smoothness; added sections III, § 5, on weak compactness and structure of normed spaces, and VII, § 4, on isomorphism to improve the norm, and Index of Citations.

Urbana, September 1972 — M. M. D.

---

* In the 3rd edition page 201.

Contents

Chapter I. Linear Spaces

§ 1. Linear Spaces and Linear Dependence

The axioms of a linear or vector space have been chosen to display some of the algebraic properties common to many classes of functions appearing frequently in analysis. Of these examples there is no doubt that the most fundamental, and earliest, examples are furnished by the n-dimensional Euclidean spaces and their vector algebras. Nearly as important, and the basic examples for most of this book, are many function spaces; for example, $C[0, 1]$, the space of real-valued continuous functions on the closed unit interval, $BV[0, 1]$, the space of functions of bounded variation on the same interval, $L^p[0, 1]$, the space of those Lebesgue measurable functions on the same interval which have summable p^{th} powers, and $A(D)$, the space of all complexvalued functions analytic in a domain D of the complex plane.

Though all these examples have further noteworthy properties, all share a common algebraic pattern which is axiomatized as follows: (Banach, p. 26; Jacobson).

Definition 1. A *linear space* L over a field Λ is a set of elements satisfying the following conditions:

(A) The set L is an Abelian group under an operation $+$; that is, $+$ is defined from $L \times L$ into L such that, for every x, y, z in L,

(a) $x+y=y+x$, (commutativity)
(b) $x+(y+z)=(x+y)+z$, (associativity)
(c) there is a w dependent on x and y such that $x+w=y$.

(B) There is an operation defined from $\Lambda \times L$ into L, symbolized by juxtaposition, such that, for λ, μ in Λ and x, y in L,

(d) $\lambda(x+y)=\lambda x+\lambda y$, (distributivity)
(e) $(\lambda+\mu)x=\lambda x+\mu x$, (distributivity)
(f) $\lambda(\mu x)=(\lambda \mu)x$,
(g) $1x=x$ (where 1 is the identity element of the field).

In this and the next section any field not of characteristic **2** will do; in the rest of the book order and distance are important, so the real field R is used throughout, with remarks about the complex case when that field can be used instead.

(1) If L is a linear space, then (a) there is a unique element 0 in L such that $x+0=0+x=x$ and $\mu 0=0x=0$ for all μ in Λ and x in L; (b) $\mu x=0$ if and only if $\mu=0$ or $x=0$; (c) for each x in L there is a unique y in L such that $x+y=y+x=0$ and $(-1)x=y$; (then for z, x in L define $z-x=z+(-1)x$ and $-x=0-x$).

(2) It can be shown by induction on the number of terms that the commutative, associative and distributive laws hold for arbitrarily large finite sets of elements; for example, $\sum_{i\leqq n} x_i$, which is defined to be $x_1+(x_2+(\cdots+x_n)\cdots)$, is independent of the order or grouping of terms in the process of addition.

Definition 2. A non-empty subset L' is called a *linear subspace* of L if L' is itself a linear space when the operations used in L' are those induced by the operations in L. If $x\neq y$, the *line through* x *and* y is the set $\{\mu x+(1-\mu)y\colon \mu\in\Lambda\}$. A non-empty subset E of L is *flat* if with each pair $x\neq y$ of its points E also contains the line through x and y.

(3) L' is a linear subspace of L if and only if for each x, y in L' and each λ in Λ, $x+y$ and λx are in L'.

Definition 3. If $E, F\subseteqq L$ and $z\in L$, define

$$E+F=\{x+y\colon x\in E \text{ and } y\in F\}. \qquad -E=\{-x\colon x\in E\},$$
$$E+z=\{x+z\colon x\in E\}, \qquad E-z=E+(-z), \qquad E-F=E+(-F).$$

(4) (a) E is flat if and only if for each x in E the set $E-x$ is a linear subspace of L. (b) The intersection of any family of linear [flat] subsets of L is linear [either empty or flat]. (c) Hence each non-empty subset E of L is contained in a smallest linear [flat] subset of L, called the *linear [flat] hull* of E.

Definition 4. If L is a linear space and $x_1, \ldots, x_n$ are points of L, a point x is a *linear combination* of these x_i if there exist $\lambda_1, \ldots, \lambda_n$ in Λ such that $x=\sum_{i\leqq n}\lambda_i x_i$. A set of points $E\subseteqq L$ is called *linearly independent* if E is not $\emptyset$ or $\{0\}$ and[1] no point of E is a linear combination of any finite subset of the other points of E. A *vector basis* (or Hamel basis) in L is a maximal linearly independent set.

(5) (a) The set of all linear combinations of all finite subsets of a set E in L is the linear hull of E. (b) E is linearly independent if and

[1] $\emptyset$ is the empty set; $\{x\}$ is the set containing the single element x.

only if for $x_1, \dots, x_n$ distinct elements of E and $\lambda_1, \dots, \lambda_n$ in Λ the condition $\sum_{i \leqq n} \lambda_i x_i = 0$ implies that $\lambda_1 = \lambda_2 = \cdots = \lambda_n = 0$.

Theorem 1. *If E is a linearly independent set in L, then there is a vector basis B of L such that $B \supseteqq E$.*

Proof. Let $\mathfrak{S}$ be the set of all linearly independent subsets S of L such that $E \subseteqq S$; let $S_1 \geqq S_2$ mean that $S_1 \supseteqq S_2$. Then if $\mathfrak{S}_0$ is a simply ordered subsystem of $\mathfrak{S}$ and S_0 is the union of all S in $\mathfrak{S}_0$, S_0 is also a linearly independent set; indeed, $x_1, \dots, x_n$ in S_0 imply that there exist S_i in $\mathfrak{S}_0$ with x_i in S_i. Since $\mathfrak{S}_0$ is simply ordered by inclusion, all x_i belong to the largest S_j and are, therefore, linearly independent. Hence $S_0 \in \mathfrak{S}$ and is an upper bound for $\mathfrak{S}_0$. Zorn's lemma now applies to assert that E is contained in a maximal element B of $\mathfrak{S}$. This B is the desired vector basis, for it is a linearly independent set and no linearly independent set is larger.

Corollary 1. *If L_0 is a linear subspace of L and B_0 is a vector basis for L_0, then L has a vector basis $B \supseteqq B_0$.*

(6) If $B = \{x_s : s \in S\}$ is a vector basis in L, each x in L has a representation $x = \sum_{s \in \sigma} \lambda_s x_s$, where σ is a finite subset of S. If $x = \sum_{s \in \sigma_1} \lambda_s x_s = \sum_{s \in \sigma_2} \mu_s x_s$, then $\lambda_s = \mu_s$ for all s in $\sigma_1 \cap \sigma_2$ and $\lambda_s = 0$ for all other s in σ_1 and $\mu_s = 0$ for all other s in σ_2. Hence each $x \neq 0$ has a unique representation in which all coefficients are non-zero, and 0 has no representation in which any coefficient is non-zero. [Also see §2, (2c).] This property characterizes bases among subsets of L.

Theorem 2. *Any two vector bases S and T of a linear space L have the same cardinal number.*

Proof. Symmetry of our assumptions and the Schroeder-Bernstein theorem on comparability of cardinals (Kelley, p. 28) show that it suffices to prove that S can be matched with a subset of T. Consider the transitively ordered system of functions Φ consisting of those functions φ such that (a) the domain $D_\varphi \subseteqq S$ and the range $R_\varphi \subseteqq T$. (b) φ is one-to-one between D_φ and R_φ. (c) $R_\varphi \cup (S \setminus D_\varphi)$ is a linearly independent set. Order Φ by: $\varphi \geqq \varphi'$ means that φ is an extension of φ'.

Every simply ordered subsystem Φ_0 of Φ has an upper bound φ_0: Define $D_{\varphi_0} = \bigcup_{\varphi \in \Phi_0} D_\varphi$ and $\varphi_0(s) = \varphi(s)$ if $s \in D_\varphi$ and $\varphi \in \Phi_0$. This φ_0 is defined and is in Φ; it is an upper bound for Φ_0. By Zorn's lemma there is a maximal φ in Φ. We wish to show that $D_\varphi = S$.

If not, then $R_\varphi \neq T$, for each s in the complement of D_φ is dependent on T but not on R_φ. If t_0 is in $T \setminus R_\varphi$, either t_0 is linearly independent of $R_\varphi \cup (S \setminus D_\varphi)$ or is dependent on it. In the former case, for arbitrary s_0 in $S \setminus D_\varphi$ the extension φ' of φ for which $\varphi'(s_0)=t_0$ has the properties (a), (b), and (c), so φ is not maximal. In the latter case, by (c) and (6)

$$t_0 = \sum_{t \in R_\varphi} \lambda_t t + \sum_{s \notin D_\varphi} \mu_s s,$$

where at least one μ_{s_0} is not zero, because t_0 is independent of R_φ. If φ' is the extension of φ for which $\varphi'(s_0)=t_0$, then φ' obviously satisfies (a) and (b); also $R_{\varphi'} \cup (S \setminus D_{\varphi'})$ is linearly independent, because otherwise t_0 would depend on $R_\varphi \cup (S \setminus D_{\varphi'})$, a possibility prevented by the choice of s_0, and again φ cannot be maximal.

This shows that if φ is maximal in Φ, then $D_\varphi = S$; then the cardinal number of S is not greater than that of T. The Schroeder-Bernstein theorem completes the proof of the theorem.

Definition 5. The cardinal number of a vector basis of L is called the *dimension* of L.

The linear space with no element but 0 is the only linear space with an empty vector basis; it is the unique linear space of dimension 0.

(7) (a) If K is the complex field and if L is a vector space over K, then L is also a vector space, which we shall call $L_{(r)}$, over the real field R. (b) The dimension of $L_{(r)}$ is twice that of L, for x and ix are linearly independent in $L_{(r)}$.

§ 2. Linear Functions and Conjugate Spaces

In this section again the nature of the field of scalars is unimportant as long as it is not of characteristic 2.

Definition 1. If L and L' are linear spaces over the same field Λ, a function F (sometimes to be called an *operator*) from L into L' is called *additive* if $F(x+y)=F(x)+F(y)$ for all x, y in L; *homogeneous* if $F(\lambda x)=\lambda F(x)$ for all λ in Λ and x in L; *linear* if both additive and homogeneous. A one-to-one linear F carrying L onto L' is an *isomorphism* of L and L'.

(1) (a) Let B be a vector basis of L and for each b in B let y_b be a point of the linear space L'. Then there is a unique linear function F from L into L' such that $F(b)=y_b$ for all b in B; precisely, using § 1, (6),

$$F\Big(\sum_{b \in \sigma} \lambda_b b\Big) = \sum_{b \in \sigma} \lambda_b y_b .$$

(b) If T_0 is a linear function defined from a linear subspace L_0 of L into a linear space L', there is an extension T of T_0 defined from L into L'.

(c) T is called *idempotent* if $TTx = Tx$ for all x in L. If L_0 is a linear subspace of L, there is an idempotent linear function (a *projection*) P from L onto L_0. (d) There is an isomorphism between L and L' if and only if these spaces have the same dimension.

Linear extension problems are much simplified by the basis theorems.

Lemma 1. *Let L and L' be linear spaces over Λ and let X be a subset of L, and let f be a function from X into L'. Then there is a linear function F from L into L' such that F is an extension of f if and only if whenever a linear combination of elements of X vanishes, then the same linear combination of the corresponding values of f also vanishes; i.e., if $\sum_i \lambda_i x_i = 0$ then $\sum_i \lambda_i f(x_i) = 0$.*

Proof. The necessity is an immediate consequence of the linearity of F. If the condition holds, define g at any point $y = \sum_i \lambda_i x_i$ in L_0, the linear hull of X, by $g(y) = \sum_j \lambda_i f(x_i)$. If also $y = \sum_j \lambda'_j x'_j$, then $\sum_i \lambda_i x_i - \sum_j \lambda'_j x'_j = 0$ so $\sum_i \lambda_i f(x_i) - \sum_j \lambda'_j f(x'_j) = 0$, and $g(y)$ is determined by y, not by its representations in terms of X. This shows at once that g is linear on L_0; (1 b) asserts that g has a linear extension F.

Definition 2. (a) If L is a linear space, then $L^{\#}$, *the conjugate space of L*, is the set of all linear functions from L into the field Λ. (b) Let S be a non-empty set of indices and for each s in S let L_s be a linear space over Λ. Let $\prod_{s \in S} L_s$ be the set of all functions x on S such that $x(s) \in L_s$ for all s in S; let $\sum_{s \in S} L_s$ be the subset of $\prod_{s \in S} L_s$ consisting of those functions x for which $\{s : x(s) \neq 0\}$ is finite. Then these function spaces are linear spaces under the definitions

$$(x + y)(s) = x(s) + y(s) \quad \text{and} \quad (\lambda x)(s) = \lambda(x(s))$$

for all x, y and all λ. They are called, respectively, the *direct product* and *direct sum* of the spaces L_s. (c) L^S is the special direct product in which all $L_s = L$.

(2) (a) $L^{\#}$ is a linear subspace of Λ^L; hence $L^{\#}$ is a linear space. (b) $\left(\sum_{s \in S} L_s\right)^{\#}$ is isomorphic to $\prod_{s \in S} (L_s^{\#})$. (c) If $\{x_s : s \in S\}$ is a basis for L and if for each s in S, f_s is the unique element of $L^{\#}$ such that $f_s(x_s) = 1$, $f_s(x_{s'}) = 0$ if $s' \neq s$, then for each x in L, $\sigma_x = \{s : f_s(x) \neq 0\}$ is a finite subset of S and for every non-empty $\sigma \supseteqq \sigma_x$, $x = \sum_{s \in \sigma} f_s(x) x_s$.

(d) If $\{x_s : s \in S\}$ is a basis in L, then L is isomorphic to $\sum_{s \in S} L_s$, where each $L_s = \Lambda$, and $L^{\#}$ is isomorphic to Λ^S. (e) If x_i, $1 \leqq i \leqq n$, are linearly independent elements of L and if λ_i, $1 \leqq i \leqq n$, are in Λ, then there exists f in $L^{\#}$ such that $f(x_i) = \lambda_i$, $1 \leqq i \leqq n$.

Definition 3. A *hyperplane* H in L is a maximal flat proper subset of L, that is, H is flat, and if $H' \supseteqq H$ and H' is flat, then $H' = H$ or $H' = L$.

(3) (a) H is a hyperplane in L if and only if H is a translation $x + L_0$ of a maximal linear proper subspace L_0 of L. (b) If $f \in L^{\#}$, if f is not 0, and if $\lambda \in \Lambda$, then $\{x : f(x) = \lambda\}$ is a hyperplane in L. (c) For each hyperplane H in L there is an $f \neq 0$, $f \in L^{\#}$, and a λ in Λ such that $H = \{x \in L : f(x) = \lambda\}$; H is linear if and only if $\lambda = 0$. (d) If the hyperplane $H = \{x : f_1(x) = \lambda_1\} = \{x : f_2(x) = \lambda_2\}$, then there exists $\mu \neq 0$ in Λ such that $f_1 = \mu f_2$ and $\lambda_1 = \mu \lambda_2$.

Definition 4. If L_0 is a linear subspace of L, define a vector structure on the *factor space* L/L_0 of all translates, $x + L_0$, of L_0 as follows: If X and Y are translates of L_0 define $X + Y$ as in §1, Def. 3 to be $\{x + y : x \in X \text{ and } y \in Y\}$; define λX to be $\{\lambda x : x \in X\}$ if $\lambda \neq 0$, $0X = L_0$. Let T_0 be the function carrying x in L to $x + L_0$ in L/L_0.

Theorem 1. *L/L_0 is a vector space and T_0 is a linear function from L onto L/L_0.*

Proof. If $x \in X$ and $y \in Y$, then $X = T_0 x = x + L_0$ and $Y = T_0 y = y + L_0$. Hence $X + Y = \{x + y + u + v : u, v \in L_0\} = \{x + y + w : w \in L_0\} = (x + y) + L_0 = T_0(x + y)$. Hence $X + Y \in L/L_0$, and T_0 is additive. Similarly $X = X + L_0$, so L_0 is the zero element of L/L_0. If $\lambda \neq 0$, then $\lambda X = \lambda T_0 x = \lambda\{x + u : u \in L_0\} = \{\lambda x + \lambda u : u \in L_0\} = \{\lambda x + v : v \in L_0\} = T_0(\lambda x)$, so $\lambda X \in L/L_0$ and $T_0(\lambda x) = \lambda T_0(x)$. If $\lambda = 0$, $0X = L_0 = T_0(0) = T_0(0x)$, so T_0 is homogeneous. Associativity, distributivity, and so on, are easily checked.

Next we improve the result of (2e).

Definition 5. A subset Γ of $L^{\#}$ is called *total over* L if $f(x) = 0$ for all f in Γ implies that $x = 0$.

Theorem 2. *Let Γ be a linear subspace of $L^{\#}$ which is total over L and let x_i, $i = 1, \ldots, n$, be linearly independent elements of L; then there exists elements f_i, $i = 1, \ldots, n$, in Γ such that $f_i(x_j) = \delta_{ij}$* (Kronecker's delta) *for $i, j = 1, \ldots, n$.*

Proof. To prove this by induction on n, begin with $n = 1$. If x_1 is a linearly independent set, then $x_1 \neq 0$; hence, by totality there is an f in Γ with $f(x_1) \neq 0$; set $f_1 = f / f(x_1)$.

Assume the result true for $n - 1$ and let $x_1, \ldots, x_n$ be independent. Then there exist $f'_1, \ldots, f'_{n-1}$ such that $f'_i(x_j) = \delta_{ij}$ for $i, j = 1, \ldots, n - 1$.

Let T map Γ into Λ^n by $(Tf)_j = f(x_j)$, $j=1,\dots,n$. We wish to show T is onto Λ^n, so we suppose, for a contradiction, that Tf is linearly dependent on the Tf'_i, $i<n$, for all f in Γ. Then $Tf = \sum_{i<n} \lambda_i Tf'_i$ so $(Tf)_j = \sum_{i<n} \lambda_i f'_i(x_j)$ for $j \leqq n$. Then for $j<n$, $f(x_j) = (Tf)_j = \lambda_j$, so $f(x_n) = (Tf)_n = \sum_{i<n} f(x_i) f'_i(x_n) = f\left(\sum_{i<n} f'_i(x_n) x_i\right)$. This yields after subtraction that $f\left(x_n - \sum_{i<n} f'_i(x_n) x_i\right) = 0$ for all f in Γ; this in turn implies that $x_n = \sum_{i<n} f'_i(x_n) x_i$, a contradiction with linear independence of the x_i. Hence there is an f in Γ such that Tf is independent of the Tf'_i, $i<n$; let $f' = f - \sum_{i<n} f(x_i) f'_i$ so $(Tf')_j = 0$ if $j<n$; let $f_n = f'/f'(x_n)$. Finally for $i<n$ let $f_i = f'_i - f'_i(x_n) f_n$. Then $f_i(x_j) = \delta_{ij}$ for $i,j \leqq n$.

Corollary 1 (Solution of equations). *If x_i, $i=1,\dots,n$, are linearly independent in L and if Γ is a linear subspace of $L^\#$ which is total over L, and if $\lambda_i \in \Lambda$, $i=1,\dots,n$, then there exists f in Γ such that $f(x_i) = \lambda_i$, $i=1,\dots,n$.*

Proof. Set $f = \sum_{i \leqq n} \lambda_i f_i$, where the f_i satisfy the conclusion of the theorem.

Corollary 2. *If $f_1,\dots,f_n$ are linearly independent elements of $L^\#$, then there exists $x_1,\dots,x_n$ in L such that $f_i(x_j) = \delta_{ij}$, and if $c_1,\dots,c_n \in \Lambda$, there is an x in L such that $f_i(x) = c_i$ for $i,j = 1,\dots,n$.*

Proof. Define Q from L into $(L^\#)^\#$ by $Qx(f) = f(x)$ for all f in $L^\#$. Then $Q(L)$ is total over $L^\#$ and Theorem 2 can be applied.

Definition 6. If E is a subset of L, define $E^\perp = \{f \in L^\# : f(x) = 0$ for all x in $E\}$. If Γ is a subset of $L^\#$, define $\Gamma_\perp = \{x \in L : f(x) = 0$ for all f in $\Gamma\}$.

Corollary 3. *Let φ be a finite subset of $L^\#$. Then the deficiency of $\varphi_\perp$, that is, the dimension of $L/\varphi_\perp$, is the number of elements in a maximal linearly independent subset of φ. Hence $(\varphi_\perp)^\perp$ is the smallest linear subspace of $L^\#$ containing φ.*

Proof. If $f_1,\dots,f_n$ is a maximal linearly independent set in φ, take $x_1,\dots,x_n$ by Cor. 2. Then if x is in L, $x - \sum_{i \leqq n} f_i(x) x_i$ is in $\varphi_\perp$. Hence the dimension of $L/\varphi_\perp$ is not greater than the number n of elements x_i. But if $X_i = x_i + \varphi_\perp$ and if $\sum_i t_i X_i = 0$, then $x = \sum_i t_i x_i$ and $t_i = f_i(x) = 0$; hence the X_i are linearly independent and $L/\varphi_\perp$ has dimension precisely n.

If f vanishes on $\varphi_\perp$, let $c_i = f(x_i)$ and $g = \sum_i c_i f_i$. Then $g - f \in \varphi_\perp{}^\perp$ and $(g-f)(x_i) = 0$, so $g - f = 0$. Hence $f = g = \sum_i c_i f_i$.

Corollary 3′. *Dually, if φ is a finite subset of L and if Γ is a linear subspace of $L^\#$ which is total over L, then the deficiency of $\varphi^\perp \cap \Gamma$ in Γ is the number of elements in a maximal linearly independent subset of φ. Hence $(\varphi^\perp \cap \Gamma)_\perp$ is the smallest linear subspace of L containing φ.*

Corollary 4. *If $H_i = \{x : f_i(x) = c_i\}$, where the f_i are linearly independent elements of $L^\#$, and if $H = \{x : f(x) = c\}$ contains $\bigcap_{i \leq n} H_i$, then there exists numbers t_i such that $f = \sum_{i \leq n} t_i f_i$ and $c = \sum_{i \leq n} t_i c_i$.*

Proof. The equations $f_i(x) = c_i$, $f(x) = c + 1$, are inconsistent; by Corollary 2 the functions $f, f_1, \ldots, f_n$ can not be linearly independent.

(4) Let T be a linear function from a linear space L into a linear space L'; let $L_0 = T^{-1}(0)$, let $L_1 = T(L)$, and let T_0 be the natural linear map of L onto L/L_0. Then L_0 and L_1 are linear subspaces of L and L', respectively, and there is an isomorphism T_1 of L/L_0 onto L_1, defined by $T_1(x + L_0) = Tx$, such that $Tx = T_1 T_0 x$ for every x in L.

Definition 7. If T is a linear function from one linear space L into another such space L', define $T^\#$, the *dual function* of T, for each f' in $L'^\#$ by $[T^\# f'](x) = f'(Tx)$ for each x in L.

(5) (a) For each f' in $L'^\#$ the function $T^\# f'$ is in $L^\#$. (b) $T^\#$ is a linear function from $L'^\#$ into $L^\#$. (c) $T^{\#-1}(0) = T(L)^\perp$, so $T^\#$ is an isomorphism of $L'^\#$ into $L^\#$ if and only if T carries L onto L'. (d) $T^\#(L'^\#) = T^{-1}(0)^\perp$, so $T^\#$ carries $L'^\#$ onto $L^\#$ if and only if T is an isomorphism of L into L'.

Theorem 3. *If L_0 is a linear subspace of L, then $L_0{}^\#$ is naturally isomorphic to $L^\#/L_0^\perp$ and $(L/L_0)^\#$ is naturally isomorphic to $L_0^\perp$.*

Proof. If i is the identity isomorphism of L_0 into L, then by (5) $U_0 = i^\#$ carries $L^\#$ onto $L_0{}^\#$ and $U_0^{-1}(0) = L_0{}^\perp$, $L_0{}^\#$ and $L^\#/L_0{}^\perp$ are isomorphic under the U_1 associated by (4) with U_0. If T_0 is the usual mapping of L onto L/L_0, then by (5) $T_0^\#$ is an isomorphism of $(L/L_0)^\#$ onto $L_0{}^\perp$.

(6) Let L be a linear space over the complex field K and for each f in $L^\#$ let $f = g + ih$, where g and h are real-valued functions, the *real* and *imaginary parts* of f. Then: (a) If $f \in L^\#$, then g and $h \in L_{(r)}{}^\#$. (See definition in § 1, (7).) (b) $h(x) = -g(ix)$. (c) The correspondence between f and g is an isomorphism of $(L^\#)_{(r)}$ and $(L_{(r)})^\#$.

(7) L and $L^\#$ (or any total linear subset Γ of $L^\#$) give examples of linear spaces in duality. L and M are said to be *dual linear spaces* if

there is a bilinear functional $\langle,\rangle$ defined on $L \times M$ such that for each $x \neq 0$ in L there is a y in M such that $\langle x, y\rangle \neq 0$, and the dual condition with L and M interchanged. Then (a) If T is defined on L by $Tx(y) = \langle x, y\rangle$ for all y in M, then each $Tx \in M^{\#}$ and the range $T(L)$ is total over M. (b) Dually, if $Uy(x) = \langle x, y\rangle$ for all x in L, then $Uy \in L^{\#}$ and $U(M)$ is total over L. (c) U is dual to T in the sense that $Tx(y) = Uy(x)$ for all x in L and y in M.

(8) Extension problems will recur again and again throughout this book. It will pay perhaps to see how simple linear extension problems are, due to the basis theorem. Let X be a linear subspace of Y, let Z be another linear space, and let f_0 be a linear function from X into Z; the problem is to find an extension f of f_0 defined and linear from Y into Z. The question for linear functions is answered by (1b), but further restrictions on the functions may make the problem insuperably difficult. It is to be noted for later use that there are several problems here, all equivalent in this linear case. These are: (a) The *"from" extension problem* in which X is fixed and Y and Z arbitrary. (b) The *"into" extension problem* in which Z is fixed and X, Y arbitrary. (c) The *projection problem* in which $X = Z$ and f_0 is the identity. Another problem which turns out ultimately to be distinct from these in most circumstances, is (d) the *subspace projection problem* in which Y is fixed, and $X = Z$ ranges over all subspaces of Y. The Hahn-Banach theorem of the next section solves an "into" extension problem: in it the range space is the reals, and, in addition to linearity, the functions are required to satisfy a domination condition.

Extension problems are considered in detail in V, § 4, VI, § 3, and VII, § 3.

§ 3. The Hahn-Banach Extension Theorem

Now we wish to consider convexity and order, so the real field R is assumed hereafter; an occasional application to complex fields is noted. A *functional* is a function with its values in the scalar field.

Definition 1. A functional p defined on a linear space L is *subadditive* if $p(x+y) \leqq p(x) + p(y)$ for all x, y in L; p is *positive-homogeneous* if $p(rx) = rp(x)$ for each $r > 0$ and each x in L; p is *sublinear* if it has both the above properties. A sublinear functional p is a *pre-norm* if $p(\lambda x) = |\lambda| p(x)$ for all λ in the field of scalars. A pre-norm p is a *norm* if $p(x) = 0$ if and only if $x = 0$.

(1) (a) If p is a sublinear functional, then $p(0) = 0$ and $-p(-x) \leqq p(x)$. (b) If p is a pre-norm in L, then $p(x) \geqq 0$ for all x in L and $\{x : p(x) = 0\}$ is a linear subspace of L.

(2) Let S be a set, let $m(S)$ be the set of all bounded real-valued functions on S, let $p_1(x)=\sup\{x(s): s\in S\}$, and let $p_2(x)=\sup\{|x(s)|: s\in S\}$. Then $m(S)$ is a linear space, p_1 is a sublinear functional in $m(S)$, and p_2 is a norm in $m(S)$. For each s_0 in $S, p_{s_0}(x)=|x(s_0)|$ is a pre-norm in $m(S)$.

Theorem 1 (Hahn-Banach Theorem). *Let p be a sublinear functional on L, let L_0 be a linear subspace of L, and let f_0 be an element of $L_0^\#$ which is dominated by p; that is, $f_0(x)\leqq p(x)$ for all x in L_0; then f_0 has an extension f in $L^\#$ which is also dominated by p.*

Proof. First we prove that f_0 has a maximal extension dominated by p. Let $\mathfrak{F}$ be the family of all linear functionals f' defined on linear subspaces L' of L such that $L_0\subseteqq L'\subseteqq L$ and f' is an extension of f_0 dominated on L' by p. Define $f'\geqq f''$ to mean that f' is an extension of f''. Then $\mathfrak{F}$ is transitively ordered, and each simply-ordered subfamily $\mathfrak{F}_0$ of $\mathfrak{F}$ has an upper bound, the f defined on the union of the domains of the f' in $\mathfrak{F}_0$ to agree with each such f' in its domain. By Zorn's lemma (Kelley, p. 33) there is a maximal f in $\mathfrak{F}$; to show that this extension has all of the properties desired, it suffices to show that its domain of definition is L. Assume then that an f' in $\mathfrak{F}$ is defined on a proper subspace L' of L; we show it can be extended.

Take z not in L' and, to discover the restrictions on any possible extension, take x, y, in L'. Then

$$f'(x)-f'(y)=f'(x-y)\leqq p(x-y)$$
$$=p(x+z+(-y-z))\leqq p(x+z)+p(-y-z),$$

so

$$-p(-y-z)-f'(y)\leqq p(x+z)-f'(x).$$

It follows that

$$\sup\{-p(-y-z)-f'(y);\, y\in L'\}\leqq \inf\{p(x+z)-f'(x);\, x\in L'\};$$

let c be any real number between these two.

In the linear space $L_1=\{x+rz: x\in L' \text{ and } r \text{ real}\}$ define f_1 by $f_1(x+rz)=f'(x)+rc$. Since each point w in L_1 determines its x and r uniquely and linearly, this defines f_1 in $L_1{}^\#$. To show f_1 dominated by p take $w=x+rz$. If $r=0$, $f_1(w)=f'(x)\leqq p(x)=p(w)$. If $r\neq 0$, by the choice of c we have for every y in L' that

$$-p(-y-z)-f'(y)\leqq c\leqq p(y+z)-f'(y).$$

Set $y=x/r$, then

$$-p\left(-\frac{x}{r}-z\right)-f'\left(\frac{x}{r}\right)\leqq c\leqq p\left(\frac{x}{r}+z\right)-f'\left(\frac{x}{r}\right).$$

Multiply by r and use the right (left) half of this if $r>0$ (if $r<0$); then

$$rc \leqq p(x+rz) - f'(x)$$

or

$$f_1(w) = f'(x) + rc \leqq p(w).$$

This extension shows f' not maximal if $L' \neq L$. Hence every maximal dominated extension of f_0 satisfies the conclusion of the theorem.

Corollary 1. *If p is sublinear on L and $x_0 \in L$, then there is an $f \in L^{\#}$ such that $f(x) \leqq p(x)$ for all x in L and $f(x_0) = p(x_0)$.*

Proof. Take $L_0 = \{r x_0 : r \text{ real}\}$ and $f_0(r x_0) = r p(x_0)$.

Definition 2. The *core* of a subset E of L is the set $\{x$: for each y in L there is an $\varepsilon(y)>0$ such that $x+ty \in E$ if $|t| < \varepsilon(y)\}$.

Geometrically speaking, this means that every line through x meets E in a set containing an interval (disc in the complex case) about x.

Definition 3. If $x, y \in L$, the *line segment* between them is the set $\{tx + (1-t)y : 0 \leqq t \leqq 1\}$. A non-empty set E in L is *convex* if for each pair of points in E the segment between them is in E. An *open segment* is a line segment minus its end-points.

(3) (a) The intersection of a family of convex sets is either empty or convex. (b) Hence each non-empty subset of a linear space L is contained in a smallest convex set, $k(E)$, *the convex hull of* E. (c) If E is a non-empty subset of L, then

$$k(E) = \left\{ \sum_{i \leqq n} t_i x_i : n = 1, 2, \ldots, x_i \in E, \quad t_i \geqq 0, \text{ and } \sum_{i \leqq n} t_i = 1 \right\}.$$

(4) Say that a set E *lies on one side of a hyperplane* H if $k(E \setminus H)$ does not intersect H. When $f \in L^{\#}$, $f \neq 0$, and $H = \{x : f(x) = c\}$, then E lies on one side of H if and only if $f(x) - c$ does not change sign in E; that is, if and only if E lies in one of the two *half-spaces* $\{x : f(x) \leqq c\}$ and $\{x : f(x) \geqq c\}$.

Lemma 1. *If p is a sublinear functional on L, if k is a positive number, and if $E = \{x : p(x) \leqq k\}$, then E is convex and the core of E is $\{x : p(x) < k\}$; hence 0 is a core point of E.*

Proof. If $x, y \in E$, then $p(tx + (1-t)y) \leqq t p(x) + (1-t) p(y) \leqq k$, so the segment from x to y is in E. If $p(x) < k$, $t \geqq 0$, and $y \in L$, $p(x + ty) \leqq p(x) + t p(y)$. If $p(y) = 0 = p(-y)$, $\varepsilon(y)$ is arbitrary; otherwise take $\varepsilon(y) = (k - p(x)) / \max[p(y), p(-y)]$.

Definition 4. Let E be a set with 0 in its core; then the *Minkowski functional* p_E is defined for each x in L by

$$p_E(x) = \inf\left\{ r : \frac{x}{r} \in E \text{ and } r > 0 \right\}.$$

Lemma 2. *If E is convex and 0 is a core point of E, then p_E, the Minkowski functional of E, is non-negative and sublinear, and p_E is a pre-norm if and only if $rE \subseteqq E$ whenever $|r|<1$.*

Proof. For each $x, x/r \in E$ if r is large enough, so $p_E(x)$, the inf of a non-empty set of positive numbers, is non-negative and finite. If $y=tx, t>0$, then

$$p_E(y) = \inf\left\{r>0: \frac{y}{r} \in E\right\} = \inf\left\{r>0: \frac{tx}{r} \in E\right\}$$

$$= \inf\left\{tr'>0: \frac{x}{r'} \in E\right\} = t\inf\left\{r'>0: \frac{x}{r'} \in E\right\} = t\,p_E(x).$$

If $x_1, x_2 \in L$, take $\varepsilon>0$ and choose r_i so that $p_E(x_i)<r_i<p_E(x_i)+\varepsilon$; then $x_i/r_i \in E$. Set $r=r_1+r_2$, then $(x_1+x_2)/r=(r_1/r)(x_1/r_1)+(r_2/r)(x_2/r_2)$ is on the segment between x_1/r_1 and x_2/r_2; by convexity, $(x_1+x_2)/r$ is in E; hence $p_E(x_1+x_2) \leqq r = r_1+r_2 < p_E(x_1)+p_E(x_2)+2\varepsilon$. Letting ε tend to 0 shows that p_E is subadditive.

Bohnenblust and Sobczyk showed that the Hahn-Banach theorem holds over the complex field:

Let L be a complex-linear space, let p be a prenorm in L, let L_0 be a complex-linear subspace of L, and let f_0 be an element of $L_0{}^{\#}$ dominated by p, in the sense that $|f_0(x)| \leqq p(x)$ for all x in L_0. Then f_0 has an extension f in $L^{\#}$ such that f is dominated by p.

Suhomlinov proved the same result for complex or quaternion scalars. Bohnenblust and Sobczyk showed that if L_0 is only real-linear, the desired conclusion may fail.

§ 4. Linear Topological Spaces

Definition 1. If a linear space L has a Hausdorff topology in which the vector operations are continuous (as functions of two variables) then L is called a *linear topological space* (LTS). If in addition every neighborhood of each point contains a convex open set containing the point, then L is called a *locally convex* linear topological space (LCS).

(1) (a) If L is an LTS and $\mathfrak{U}$ is a neighborhood basis of 0, then $\mathfrak{U}_x = \{U+x: U \in \mathfrak{U}\}$ is a neighborhood basis at x. (b) Hence every LTS has a uniform structure compatible with its topology and vector structure, and must be a completely regular space [Kelley, Chapter 6].

(2) (Von Neumann [2], Wehausen). If L is an LTS, it has a neighborhood basis $\mathfrak{U}$ at 0 such that (a) 0 is the only point common to all U in $\mathfrak{U}$; (b) if $U, V \in \mathfrak{U}$, then there is a W in $\mathfrak{U}$ such that $W \subseteqq U \cap V$;

(c) if $U \in \mathfrak{U}$ and $|r| \leqq 1$, then[2] $rU \subseteqq U$; (d) if $U \in \mathfrak{U}$ there exists $V \in \mathfrak{U}$ such that $V+V \subseteqq U$; (e) 0 is a core point of each U in $\mathfrak{U}$. L is also locally convex, if and only if $\mathfrak{U}$ can also be chosen so that (f) every U in $\mathfrak{U}$ is convex. Conversely, if a neighborhood basis at 0 is chosen to satisfy (a)—(e), and neighborhoods of other points are defined [as in (1a)] by translations of the neighborhood system at 0, then L becomes an LTS, which is locally convex if (f) also holds. Finally, (c) and (d) imply (g) for every U in $\mathfrak{U}$ and $k>0$ there exists V such that $rV \subseteqq U$ if $|r| \leqq k$.

(3) (a) Any linear subset of an LTS becomes an LTS under the relative topology [Kelley, p. 51] determined from L. (b) With the product-space topology [Kelley, p. 90] in which a neighborhood basis $\mathfrak{U}$ of 0 in R^S is the set of all $U(\sigma,\varepsilon)=\{x: |x(s)|<\varepsilon$ for each s in $\sigma\}$, with $\varepsilon>0$ and σ a finite subset of S, the space R^S is an LCS. (c) If L is a linear space, then $L^{\#}$ is a closed subspace of R^L.

(4) Let L be an LTS and let X, Y be subsets of L. (a) If X is open and $r \neq 0$, then rX is open. (b) If X or Y is open, $X+Y$ is open. (c) If X is open, so is the convex hull of X. (d) The interior of a convex set is convex or empty. (e) If X is closed and Y is compact, then $X+Y$ is closed.

Lemma 1. *If f from one* LTS *L to another L' is additive and continuous at* 0, *then f is uniformly continuous and real-homogeneous.*

Proof. Let x_0 be a point of L; then $U'+f(x_0)$ is a neighborhood of $f(x_0)$ if and only if U' is a neighborhood of 0 in L'. Then there is a neighborhood U of 0 in L such that $f(U) \subseteqq U'$. Hence $f(x_0+U)=f(x_0)+f(U)$ is contained in $f(x_0)+U'$; i.e., f is continuous at every x_0 if it is continuous at 0. This proof gives uniform continuity as an extra bonus with no more work. In any linear space an additive function is homogeneous over the rational field; this is proved (i) by induction for integers, (ii) by change of variable for reciprocals of integers, and (iii) by combining these for arbitrary rationals. Then if r is real and (r_n) is a sequence of rationals converging to r, continuity of multiplication implies $(r_n x)$ converges to rx. Hence

$$f(rx) = \lim_{n \in \omega} f(r_n x) = \lim_n r_n f(x) = \left(\lim_n r_n\right) f(x) = r f(x).$$

Corollary 1. *An additive functional is continuous if and only if it is bounded on some open set in L.*

---

[2] A set U with this property is called *symmetric star-shaped* if the field is real, *discoid* if the field is complex.

Proof. If f is continuous, $f^{-1}((-1,1))$ is open and f is bounded on it. If f is bounded on an open set U by a number k, and if $x_0 \in U$, then f is bounded by $2k$ on $U - x_0$, which contains a U_1 in $\mathfrak{U}$. By (2g) f is continuous at 0; the lemma asserts it is continuous everywhere.

(5) A sublinear functional p is continuous if and only if it is bounded on an open set and if and only if $\{x: p(x) < 1\}$ is open, and if and only if p is continuous at 0.

Corollary 2. *A linear functional f on L is continuous if and only if there is an open set U in L and a value t which f does not take in U. Hence an f in $L^{\#}$ is continuous if and only if $L_0 = f^{-1}(0)$ is closed.*

Proof. By translation of U and additivity of f it can be assumed that $0 \in U$; by (2), U contains a neighborhood V of 0 such that $rV \subseteqq V$ if $|r| \leqq 1$. Then $v \in V$ and $|r| \leqq 1$ imply that $rf(v) = f(rv) \neq t$; that is $f(v) \neq t/r$ if $|r| \leqq 1$. Hence $|f(v)| < |t|$ if $v \in V$; Corollary 1 asserts that f is continuous.

Lemma 2. *Every line in an* LTS *L is uniformly homeomorphic to the real number system R; more precisely, for each $x \neq 0$ in L the mapping $f(r) = rx$ is a uniformly bicontinuous one-to-one linear function from R onto R_1, the line through 0 and x.*

Proof. f is linear and one-to-one. Continuity of f and of f^{-1} follows from Corollary 2.

Definition 2. An *isomorphism* T of one LTS L *into* another LTS L' is an algebraic isomorphism (Def. 2,1) of L onto a linear subspace L_0 of L' such that T and T^{-1} are both continuous. L and L' are called *isomorphic* whenever there is an isomorphism of L *onto* L'.

(6) (von Neumann [2]) (a) An open subset U of an LTS L is convex if and only if (2f′) $U + U = 2U$. (b) If a subset U of L satisfies (2f′), then so does the interior of U. (c) An LTS is locally convex if and only if there exists a neighborhood basis at zero consisting of sets satisfying (2a to e) and (2f′).

Lemma 3. *Every one-dimensional subspace H of an* LTS *L is closed in L.*

Proof. Suppose $(x_n, n \in \Delta)$ is a net in H such that there is an x in L for which $\lim_n x_n = x$. Then (x_n) is a Cauchy net in H; if $z \neq 0$ is in H, the (uniformly) continuous transformation $t \leftrightarrow tz$ between H and R carries $(x_n) = (t_n z)$ into a Cauchy net (t_n) in R. Since R is a complete metric space, (t_n) converges to some limit t. Then $tz = \lim_n t_n z = \lim_n x_n = x$, so $x \in H$.

Corollary 3. *Let L be an* LTS *and let L_0 be a closed subspace. Let L_1 be a line in L which meets L_0 only at* 0 *and let $L_2 = L_1 + L_0$. Then* (i) *L_2 is closed in L, and* (ii) *if $0 \neq x \in L_1$, the natural correspondence $(y,r) \leftrightarrow y + rx$ between L_2 and $L_0 \times R$ is a homeomorphism.*

Proof. In L_2 define f by $f(z) = f(y+rx) = r$, where $r \in R$ and $y \in L_0$; then f is linear and $f^{-1}(0) = L_0$, which is closed in L_2 since it is closed in L. By Cor. 2, f is continuous; hence for $z = y + rx$, r and y are continuous linear functions of z. Therefore, the function F defined by $F(y+rx) = (y,r)$ is a continuous function from L_2 onto $L_0 \times R$. Continuity of the vector operations asserts that F^{-1} is continuous. Hence F is a homeomorphism.

If $(z_n, n \in \Delta)$ is a net in L_2 and if $\lim_n z_n = z \in L$, then $z_n = y_n + r_n x$, so $z_n - z_m = (y_n - y_m) + (r_n - r_m) x \to 0$. By continuity of f, $r_n - r_m \to 0$; since R is complete, there is an r such that $\lim_n r_n = r$. Hence $\lim_n r_n x = rx$, and $\lim_n y_n = z - rx$. But L_0 is closed, so $y = z - rx \in L_0$; hence $z = y + rx \in L_2$ and L_2 is closed.

Theorem 1. *If L_0 and L_1 are linear subspaces of an* LTS *L such that L_0 is closed and L_1 is finite dimensional, then $L_0 + L_1$ is closed in L; if also $L_0 \cap L_1 = \{0\}$ then $L_0 + L_1$ has the topology of $L_0 \times L_1$; that is, the algebraic isomorphism between these spaces is bicontinuous.*

This is proved from Corollary 3 by induction on the dimension of L_1.

Corollary 4. (a) *Every finite-dimensional subspace of a linear top. space is closed.* (b) *Every linear functional on a finite-dimensional* LTS *is continuous.* (c) *Every linear function from a finite-dimensional* LTS *into any* LTS *is continuous.* (d) *Every algebraic isomorphism of one finite-dimensional* LTS *onto another is a homeomorphism.*

(7) If L_0 is a closed linear subspace of an LTS L, topologize the factor space $L_1 = L/L_0$ by making T_0, the natural mapping of L onto L_1, interior and continuous, that is, let U_1 in L_1 be open if and only if $U = T_0^{-1}(U_1) = \bigcup_{X \in U_1} X$ is open in L. Then L_1 is also an LTS. L_1 is locally convex if L is.

Definition 3. If L is an LTS, let L^*, the *conjugate space* of L, be the set of those functions in $L^\#$ which are continuous.

Theorem 2. *If L is an* LCS, *then L^* is total over L.*

Proof. If $0 \neq x_0 \in L$, there is a convex symmetric (or discoid) neighborhood U of 0 such that $x_0 \notin U$; let p be the Minkowski functional of U. Then $p(x_0) \geqq 1$, p is a pre-norm, and Cor. 3,1 asserts that there is f with $f(x_0) = p(x_0)$ and $f(x) \leqq p(x)$ for all x in L. Since $-p(x) = -p(-x)$

$\leqq f(x) \leqq p(x)$, $|f(x)|$ is bounded on U by 1; Corollary 1 asserts that f is continuous.

This prepares for one of the simple appearances of the projection or extension problems.

Theorem 3. *If n is an integer and L_n is an n-dimensional subspace of an* LCS L, *then* (a) *every linear function from L_n into an* LTS L' *has a linear continuous extension defined on all of L and* (b) *there is a linear, continuous, idempotent mapping (a projection) of L onto L_n.* (c) *If L_0 is a linear subspace of an* LCS L, *every linear continuous function from L_0 into L_n has a linear continuous extension defined on all of L into L_n.*

Proof. For (b), Take $x_1, \ldots, x_n$ a basis for L_n and use Theorem 2,2 with $\Gamma = L^*$ to get f_i in L^* with $f_i(x_j) = \delta_{ij}$. Set $Px = \sum_{i \leqq n} f_i(x) x_i$ for all x in L. For (a) any linear f_0 on L_n is continuous; set $f(x) = f_0(Px)$. For (c) observe that L_n is isomorphic to the linear topological space R^n, so a linear continuous f from L_0 to L_n determines n linear continuous functionals on L_0; if each f_i has an extension F_i in L^*, the F with these components is a continuous linear map of L into L_n which is an extension of f. But if f_i is a linear continuous functional defined on L_0, let U be a convex neighborhood of 0 for which $|f_i(x)| \leqq 1$ if $x \in U \cap L_0$. Use the Hahn-Banach theorem with p the Minkowski functional of U to extend f_i to an F_i linear and bounded on U by 1. Cor. 1 says that F_i is continuous.

(8) (a) $L^* \neq \{0\}$ if and only if there is in L a convex open proper subset. (b) If L^* is total over L, an LTS, then (a) and (b) of Theorem 3 hold.

Definition 4. A subset X of an LTS is called *bounded* if for every neighborhood U of 0 there is an $n > 0$ such that $nU \supseteqq X$. X is *totally bounded* (or *precompact*) if for every U there is a finite set $x_1, \ldots, x_n$ in L such that $\bigcup_{i \leqq n} (x_i + U) \supseteqq X$.

Banach gave the following characterization of bounded sets in a linear metric space; see Hyers for the general case, and II, 3, (12) for Mackey's criterion.

Lemma 4. *X is a bounded subset of an* LTS L *if and only if for every sequence $(x_n, n \in \omega)$ in X and every sequence $(\lambda_n, n \in \omega)$ of non-negative numbers converging to zero the sequence $(\lambda_n x_n)$ converges to zero in L.*

Proof. If X is bounded, if $x_n \in X$, and if $\lambda_n \to 0$, then for each neighborhood U of 0 there is a $K > 0$ such that $\lambda X \subseteqq U$ if $\lambda < K$. Take n_U

so large that $\lambda_n < K$ when $n > n_U$; then $\lambda_n X \subseteqq U$ so $\lambda_n x_n \in U$ if $n > n_U$; that is $\lambda_n x_n \to 0$. If X is not bounded, there exists U in $\mathfrak{U}$ such that for no n is X/n contained in U; that is, for each n there is x_n in X such that $x_n/n \notin U$. Take $\lambda_n = 1/n$ to have $\lambda_n \to 0$ but $\lambda_n x_n$ not tending to zero.

(9) (a) If X is [totally] bounded, so are scalar multiples of X and the closure of X. (b) If X_i are [totally] bounded, so are $X_1 \cup X_2$ and $X_1 + X_2$. (c) If L is locally convex and X is [totally] bounded, then so is the convex hull of X. [See (17b).]

(10) If T is a continuous linear operator from one LTS L to another L', then T carries [totally] bounded sets in L to [totally] bounded sets in L'.

(11) (a) Every bounded closed subset of a finite-dimensional LTS is compact. (b) Every compact subset of an LTS is totally bounded. (c) Every totally bounded subset of an LTS is bounded. (d) In an LCS a set E is totally bounded if and only if for every convex symmetric neighborhood U of 0 there is a finite set φ such that E is contained in $k(\varphi \cup U)$, the convex hull of φ and U. [If $\varphi/2 + U/2 \supseteqq E$, then $k(\varphi \cup U) \supseteqq E$. If $k(\varphi \cup U/2) \supseteqq E$, then there is a finite set ψ such that $k(\varphi) \subseteqq \psi + U/2$, so $E \subseteqq k(\varphi \cup U/2) \subseteqq \psi + U$.]

(12) A linear topological space L is called *locally bounded* if there exists a bounded open set, *metrisable* if there is a metric invariant under translations which gives the original topology of L, *normable* if there is a norm $\|\ldots\|$ in L such that the metric $\|x-y\|$ determines the topology of the space. (a) L is metrisable if and only if it satisfies Hausdorff's first denumerability condition. (See Kelley for the pertinent theory of metrisation of uniform structures.) (b) If L is locally bounded, then it is first-denumerable. (c) L is normable if and only if it is both locally convex and locally bounded [Kolmogoroff].

(13) Except for a few examples, we shall be concerned with locally convex linear topological spaces. It is worth noting that each linear space L determines a strongest topology making L into an LCS. Let $\mathfrak{U}$ be the family of all sets satisfying the conditions (2c, e, f); that is, this neighborhood basis consists of all convex, symmetric sets with 0 as a core point. Then this family satisfies the other conditions of (2) and yields the strongest locally convex topology which makes L an LTS; that is, if more open sets are introduced either the space is no longer LC or the vector operations are no longer continuous. (For example, the discrete topology is "locally convex", but multiplication by scalars is not continuous.) This built-in topology we shall call the *convex core* topology of L. Now let L be a linear space with the convex core topology and let L_0 be a linear subspace of L. (a) The relative topology imposed on L_0 is the convex core topology of L_0. (b) L_0 is

closed in L. (c) In $L_1 = L/L_0$ the quotient-space topology imposed by L is the convex core topology of L_1.

(14) If L is a linear space with the convex core topology, then (a) $L^* = L^\#$; indeed, (b) every linear function from L into an LCS L' is continuous, and (c) every sublinear functional on L is continuous.

(15) If L is an LTS, then every interior point of a set E is a core point of E.

(16) In the convex core topology every flat set is closed and every core point of a convex set is an interior point.

(17) In every LTS the uniform structure can be defined by a net of pre-metrics. (See Kelley for the corresponding theory in uniform spaces.) In an LCS L the topology can be defined by a net $(p_U, U \in \mathfrak{U})$ of continuous pre-norms, where $\mathfrak{U}$ is a cofinal subset of the directed system of all neighborhoods of 0 in L such that each U in $\mathfrak{U}$ is convex and symmetric (or discoid). (a) If L is a linear space, the convex core topology is defined by the net of all pre-norms in L, ordered by $p \geqq q$ if and only if $p(x) \geqq q(x)$ for all x, or by its cofinal subset of all norms in L. (b) A set X in an LCS L is bounded if and only if every continuous pre-norm in L is bounded on X. (c) Each LCS is isomorphic to a linear subspace of a product of normed spaces (II, 3, (11)) [Grothendieck (1)].

(18) Under the correspondence of § 2, (6), $(L_{(r)})^*$ is isomorphic to $(L^*)_{(r)}$.

§ 5. Conjugate Spaces

If L and L' are linear topological spaces, it follows directly from the linearity and continuity of the composite of two linear continuous functions that *if T is a linear continuous function from L into L', then $T^\#$ carries L'^* into L^**. Hence we can define T^*, the *conjugate* or *adjoint* of T, from L'^* into L^* to be the restriction of $T^\#$ to L'^*. To discuss continuity of T^* requires topologies in L^* and L'^*. Because $L^* \subseteqq L^\# \subseteqq R^L$ the product topology in R^L imposes a relative topology on L^*; we call this the *weak**- or *w*-topology* of L^*. It is the simplest to define of the large family of topologies which are described in Theorem 1 below. $w^*(E)$ is used for the weak*-closure of E.

Definition 1. If $E \subseteqq L$, then E^π, the *polar set* of E, is the subset of L^* defined by $E^\pi = \{f : f(x) \geqq -1 \text{ for each } x \text{ in } E\} = \bigcap_{x \in E} \{f : f(x) \geqq -1\}$. If $F \subseteqq L^*$, then $F_\pi = Q^{-1}(F^\pi) = \bigcap_{f \in F} \{x : f(x) \geq -1\}$.

(1) (a) E^π is convex, w^*-closed, and contains 0. (b) If $0 \leqq r \leqq 1$, then $r(E^\pi) \subseteqq E^\pi$. (c) If $r > 0$, then $(rE)^\pi = E^\pi/r$. (d) If $E_1 \subseteqq E_2$, then $E_1^\pi \supseteqq E_2^\pi$. (e) If $K(E)$ = closed convex hull of E, then $K(E)^\pi = E^\pi$

$=K(E\cup\{0\})^\pi$. (f) $(E_1\cup E_2)^\pi=E_1^\pi\cap E_2^\pi$. (g) If E is symmetric or linear, so is E^π.

These are many of the properties of neighborhoods of 0 in L^*; to get topologies in L^* we choose suitable families of sets in L.

Definition 2. A family $\mathfrak{E}$ of subsets of an LTS L is called a *topologizing family for* L^* if (a) each finite subset φ of L is contained in at least one E of $\mathfrak{E}$. (b) If E_1 and $E_2\in\mathfrak{E}$, then there is an E_3 in $\mathfrak{E}$ such that $E_3\supseteqq E_1\cup E_2$. (c) If $E_1\in\mathfrak{E}$, there exists $E_2\in\mathfrak{E}$ such that $E_2\supseteqq 2E_1$. (d) Each E in $\mathfrak{E}$ is a symmetric bounded subset of L.

(2) Examples of families satisfying these conditions are: (a) The set $\mathfrak{F}$ of all finite symmetric subsets of L. (b) The set $\mathfrak{K}$ of all symmetric compact convex subsets of L. (c) The set $\mathfrak{P}$ of all totally bounded symmetric subsets of L. (d) The set $\mathfrak{B}$ of all bounded symmetric subsets of L.

Theorem 1. *If $\mathfrak{E}$ is a topologizing family for L^* and if $\mathfrak{E}^\pi$ is the family of polar sets $\{E^\pi\colon E\in\mathfrak{E}\}$, then $\mathfrak{E}^\pi$ satisfies the von Neumann conditions* [§ 4, (2)] *for a locally convex neighborhood basis at* 0 *and therefore determines a topology in L^* which makes L^* an* LCS. [When it is pertinent to distinguish the topology under discussion, this LCS will be denoted by $(L^*,\mathfrak{E}^\pi)$.]

Proof. Refer to the properties of (2) of § 4, § 4, (2c, f) follow from (1b, a). § 4, (2a) comes from Def. 2, (a). § 4, (2b) comes from Def. 2, (b) and (1f). § 4, (2d) follows from § 4, (2f'), (1c), and Def. 2, (c). § 4, (2e) comes from Def. 2, (d).

Corollary 1. $(L^*,\mathfrak{F}^\pi)$ *(which is also* (L^*,w^*)*),* $(L^*,\mathfrak{K}^\pi)$, $(L^*,\mathfrak{P}^\pi)$, *and* $(L^*,\mathfrak{B}^\pi)$ *are all locally convex spaces; each of the last three has a stronger topology than its predecessor in the list.*

Corollary 2. *If $\mathfrak{E}$ is a topologizing family for L^*, then the $\mathfrak{E}^\pi$ topology is not weaker than the $\mathfrak{F}^\pi$ nor stronger than the $\mathfrak{B}^\pi$ topology.*

Corollary 3. *If $\mathfrak{E}$ is a topologizing family for L^* and if for each x in L, Qx is that element of $L^{*\#}$ defined by*

$$Qx(f)=f(x)\quad \text{for all } f \text{ in } L^*,$$

then $Qx\in(L^*,\mathfrak{E}^\pi)^*$.

Proof. It suffices by Cor. 2 to prove $Qx\in(L^*,\mathfrak{F}^\pi)^*$. But $[\{-x\}\cup\{x\}]^\pi$ is an $\mathfrak{F}^\pi$-neighborhood of 0 and Qx is bounded on it. Qx is obviously linear, so Cor. 4,1 asserts that Qx is $\mathfrak{F}^\pi$-continuous; that is, $Qx\in(L^*,\mathfrak{F}^\pi)^*$.

Corollary 4. *If $\mathfrak{E}$ is a topologizing family for L^*, then Q is linear from L into $(L^*,\mathfrak{E}^\pi)^*$; L^* is total over L if and only if Q is one-to-one between L and $Q(L)$.*

Theorem 2. *If T is a continuous linear operator from one* LTS L *into another L', then T^* is a continuous linear operator from L'^* into L^* if in each of L'^*, L^* the same one of the topologies $\mathfrak{F}^\pi$, $\mathfrak{K}^\pi$, $\mathfrak{P}^\pi$, or $\mathfrak{B}^\pi$ is used.*

The proof depends only on showing that T carries each symmetric and finite, or convex compact, or totally bounded, or bounded set of L to the same kind of set in L'.

Lemma 1. *If L^* is total over L, let φ be a finite subset of L and let ξ be an element of $L^{*\#}$ which vanishes on $H=\varphi^\perp\cap L^*$. Then ξ is a linear combination of the set $Q\varphi$, so it is in $Q(L)$.*

Proof. Choose $x_1,\ldots,x_n$ a maximal linearly independent subset of φ and then take $f_1,\ldots,f_n$ in L^* by Th. 2,2 to get $f_i(x_j)=\delta_{ij}$. Define P on L^* by $Pf=\sum_{i\leqq n} f(x_i)f_i$; then $f-Pf$ vanishes on φ for all f in L^*, so $\xi(f-Pf)=0$ for all f in L^*, or

$$\begin{aligned}\xi(f)=\xi(Pf)=\xi\Big(\sum_{i\leqq n} f(x_i)f_i\Big)&=\sum_{i\leqq n} f(x_i)\xi(f_i)=\sum_{i\leqq n}\xi(f_i)Q\,x_i(f)\\ &=\Big(\sum_{i\leqq n}\xi(f_i)Q\,x_i\Big)(f).\end{aligned}$$

Hence $\xi=\sum_{i\leqq n}\xi(f_i)Q\,x_i$; that is, ξ is a linear combination of elements of $Q\varphi$; since Q is linear, $\xi=Q\Big(\sum_{i\leqq n}\xi(f_i)\,x_i\Big)\in Q(L)$.

Theorem 3. *If L^* is total over L, Q is an algebraic isomorphism of L onto $(L^*,w^*)^*$.*

Proof. If $\xi\in(L^*,w^*)^*$, then there is a finite symmetric set φ in L such that ξ is bounded on φ^π; then $H=\varphi^\perp\cap L^*\subseteqq\varphi^\pi$, so $|\xi(f)|\leqq K$ on H. But $tf\in H$ if $f\in H$ and $t\in R$, so $|t\,\xi(f)|\leqq K$ if $f\in H$ and $t\in R$; hence $\xi(f)=0$ if $f\in H$. By the lemma $\xi\in Q(L)$; the other properties of Q follow from Corollary 4.

Lemma 2. *If Γ is a linear subset of $L^\#$ which is total over L, then Γ is w^*-dense in $L^\#$.*

Proof. If $f\in L^\#$ and φ is a finite symmetric subset of L, let $x_1,\ldots,x_n$ be a maximal linearly independent subset of φ. Th. 2,2 gives a set $f_1,\ldots,f_n$ of elements of Γ such that $f_i(x_j)=\delta_{ij}$, $i,j\leqq n$. Therefore $f-\sum_{i\leqq n} f(x_i)f_i\in\varphi^\perp\subseteqq\varphi^\pi$; that is, $f+\varphi^\pi$ contains an element of Γ.

Theorem 4. *If $\mathfrak{E}$ is a topologizing family of L^* and if L^* is total over L, then Q is an algebraic isomorphism of L onto a weak*-dense subset of $(L^*,\mathfrak{E}^\pi)^*$; indeed, $Q(L)$ is w^*-dense in $L^{*\#}$.*

Proof. Increasing $\mathfrak{E}$ increases $(L^*, \mathfrak{E}^\pi)^*$, so $Q(L) = (L^*, w^*)^* \subseteqq (L^*, \mathfrak{E}^\pi)^*$. Lemma 2 shows that $Q(L)$ is weak*-dense in $L^{*\#}$.

Definition 3. If $E \subseteqq L^*$, let $E_\pi = \{x : x \in L \text{ and } f(x) \geqq -1 \text{ for all } f \text{ in } E\}$. The *weak topology of* L has its neighborhoods of 0 given by the sets ψ_π, where ψ runs over finite symmetric subsets of L^*. $w(E)$ is used for the weak closure of E.

(3) (a) The weak topology makes L an LCS if and only if L^* is total over L; if L^* is not total over L, all that fails is the separation axiom. (b) If L^* is total over L (for example, if L is an LCS), then the w-topology of L is the image under Q^{-1} of the w^*-topology of $Q(L) \subseteqq L^{*\#}$.

(4) Use the hypotheses of Cor. 3; then (a) Q is a linear function from L onto a weak*-dense subset of $L^{*\#}$. (b) Q is continuous from L (even (L, w)) onto $(L^*, w^*)^*$, and if L^* is total, Q is an algebraic and topological isomorphism between (L, w) and $((L^*, w^*)^*, w^*)$. (c) If L is locally convex, then Q^{-1} is continuous from $Q(L)$, provided with the topology defined from the determining family $\mathfrak{E}'$ of w^*-bounded sets in L^*, onto L.

Lemma 3. *If* U *is a neighborhood of* 0 *in an* LTS, *then* U^π *is* w^*-*compact.*

Proof. U may be assumed convex without changing U^π; then let p be the Minkowski functional of U, for each x in L let I_x be the closed interval $[-p(x), p(-x)]$, and let $X = \prod_{x \in L} I_x$. Then X is compact by Tychonoff's theorem and U^π is the set of all linear functions in X. § 4, (3c) shows that a w^*-limit of linear functions is itself linear, so U^π is closed in the compact set X. Therefore U^π is itself w^*-compact.

(5) If L and L' are linear spaces in duality, [§ 2, (7)], say that *topologies for* L *and* L' *are consistent with the duality* if the images $T(L)$ and $U(L')$ in $L'^{\#}$ and $L^{\#}$ are precisely the sets of functionals continuous in the given topologies. Define the polar set E^π for E a subset of L to be $\{y : \langle x, y\rangle \geqq -1 \text{ if } x \in E\}$, and dually for subsets of L'. (a) There is a weakest locally convex topology, that given in L and in L' by the polars of finite sets, consistent with a given duality. (b) The polar sets of convex, w-compact sets in L (in L') determine the strongest topology in L' (in L) consistent with the given duality. [Bourbaki, [2]. Ch. IV, § 2, No. 3.] (c) The $\mathfrak{B}^\pi$ topology in L^* is not always consistent with the duality between L and L^*, for $(L^*, \mathfrak{B}^\pi)^*$ may be much larger than $Q(L)$. [See Chapter II, 2, (6).] [Bourbaki discusses duality at the length suited to its great importance in linear topological spaces.]

(6) If T is a continuous linear function from one LTS L into another L' and if Q and Q' are the natural mappings of L and L' into $L^{*\#}$ and $L'^{*\#}$, then $Q'T = T^{*\#}Q$. (This is improved for normed spaces in Cor. II, 5,5.)

(7) If T is a continuous linear operator from one LCS L into another L', then $T^*(L'^*)$ is w^*-dense in $T^{-1}(0)^\perp \cap L^*$.

(8) From Mazur's theorem of the next section and from duality it can be proved that if L is an LCS, then for each subset E of L [of L^*] the set $E^\pi{}_\pi [E_\pi{}^\pi]$ is the smallest $w-[w^*-]$ closed convex set containing E and 0. Similarly $E^\perp{}_\perp [E_\perp{}^\perp]$ is the smallest $w-[w^*-]$ closed linear set containing E.

(9) (a) Continuing 2, (2b) we have for each index set S and each LTS L_s, $s\in S$, that when the product topology is used in $\prod_{s\in S} L_s$, then $\left(\prod_{s\in S} L_s\right)^*$ is algebraically isomorphic to $\sum_{s\in S}(L_s^*)$. (b) Hence the weak topology of the product space is the product of the weak topologies.

§ 6. Cones, Wedges, Order Relations

Definition 1. A *wedge* W in a real-linear space L is a convex set such that $tW\subseteqq W$ for every $t\geqq 0$. A *cone* is a wedge which contains no line through 0.

(1) If $f\in L^\#$, the half space $\{x: f(x)\geqq 0\}$ is a wedge but is not a cone. A wedge is a cone if and only if $x\in W$ and $-x\in W$ imply $x=0$.

Definition 2. An *ordered linear space* (OLS) is a linear space L in which some wedge W, called the *positive wedge*, has been assigned. Then the (partial) order relation in L is determined from: $x\geqq y$ (or $y\leqq x$) whenever $x-y\in W$.

(2) (a) $\geqq$ is reflexive and transitive. (b) Translation and multiplication by positive numbers preserve order; multiplication by negative numbers reverses order. (c) $x\geqq y$ and $y\geqq x$ imply $y=x$ if and only if the wedge defining the order is a cone. (d) $x\geqq 0$ if and only if $x\in W$.

(3) Conversely, if an order relation is given in a linear space satisfying (2, a and b), then (2d) defines a wedge (a cone if the antisymmetry condition of (2c) holds) such that the order determined from the wedge is the given order.

Definition 3. If W_i is a wedge in a linear space L_i, $i=1,2$, and if T is a function from L_1 into L_2, T is called *non-negative* if $T(W_1)\subseteqq W_2$; T is called *monotone* if $x\geqq_1 y$ implies $Tx\geqq_2 Ty$.

(4) An additive function from L_1 into L_2 is non-negative if and only if it is monotone.

Theorem 1 (The monotone extension theorem.) *Let L be a linear space with a wedge W; let L_0 be a linear subspace of L such that for each x in L, $x+L_0$ meets W if and only if $-x+L_0$ meets W; let f_0*

in $L_0^\#$ *be monotone. Then there is an extension* f *of* f_0 *which is monotone and in* $L^\#$.

Proof. An application of Zorn's lemma shows that there is a maximal monotone linear extension f_1 of f_0 defined on some linear subspace L_1 of L. To prove $L_1 = L$ we observe two things.

(a) L_1 inherits from L_0 its special relation with W.

Take x such that $x + L_1$ meets W; that is, there is a y in L_1 such that $x - y \in W$. Then $x - y \in x - y + L_0$, so $y - x + L_0$ meets W also; that is, there is a z in L_0 such that $y - x - z \in W$; but $y - z \in L_1$ so $-x + L_1$ meets W.

(b) If $L_1 \neq L$, then f_1 has a proper extension f_2. If possible, take x not in L_1 and let $A = \{y : y \in L_1 \text{ and } x \geqq y\}$. If A is not empty (a) shows that $B = \{z : z \in L_1 \text{ and } z \geqq x\}$ is not empty either. Set $a = \sup\{f_1(y) : y \in A\}$ and $b = \inf\{f_1(z) : z \in B\}$. Since $z \geqq x \geqq y$, $f_1(z) \geqq f_1(y)$ for all y in A, z in B, so $a \leqq b$; let c be a number between a and b, let $L_2 = \{u + tx : u \in L_1$ and t real$\}$, and define f_2 in L_2 by $f_2(u + tx) = f_1(u) + tc$.

In case A is empty, so is B; let c be any real number and then define f_2 as above. In either case f_2 is non-negative and linear and extends f_1. Hence $L_1 = L$ if f_1 is maximal; any such maximal extension of f has the desired properties.

Corollary 1 [Kreĭn-Rutman, Th. 1.1.] *Let* L *be a linear topological space with a wedge* W *and let* f_0 *be a monotone linear functional defined on a linear subspace* L_0 *which contains an interior point of* W*; then* f_0 *has a monotone extension in* L^*.

Proof. Under this hypothesis on L_0 every translation of L_0 meets W. Continuity of f follows from Cor. 4, 2.

(5) The monotone extension theorem can also be proved as an application of the Hahn-Banach theorem. Set $p(x) = \inf\{f_0(y) : y \in L_0$ and $y \geqq x\}$ if x is in the linear hull of W and L_0. The extension to the rest of L is made by Lemma 2,1.

Another important consequence of the Hahn-Banach theorem is

Theorem 2 (Mazur's theorem. Ascoli, Mazur [1], Bourgin [1].) *Let* K *be a convex set with non-empty interior in a* LTS L *and let* E *be a flat subset of* L *which contains no interior point of* K*. Then there is a closed hyperplane* H *such that* H *contains no interior point of* K *but* $H \supseteqq E$*; in other words, there exists* f *in* L^* *and a real number* c *such that* $f(x) = c$ *for all* x *in* E *and* $f(y) < c$ *for all* y *interior to* K.

Proof. By translation we can suppose that 0 is an interior point of K; let L_0 be the smallest linear subspace of L containing E. Then E is a hyperplane in L_0, so there exists f_0 in $L_0^\#$ such that $E = \{x : f_0(x) = 1\}$.

Let p be the Minkowski functional of K; the assertion that E contains no point of the interior of K, $\{x: p(x)<1\}$, says that $f_0(x)=1 \leqq p(x)$ if $x \in E$. By homogeneity $f_0(tx) \leqq p(tx)$ if $x \in E$ and $t>0$, while $f_0(tx) \leqq 0 \leqq p(tx)$ if $t \leqq 0$. Hence f_0 is dominated by p. By the Hahn-Banach theorem f_0 has an extension f in $L^{\#}$ which is also dominated by p. Let $H=\{x: f(x)=1\}$; then $f(y) \leqq p(y)<1$ in the interior of K, so, by Cor. 4,1, f is continuous and H is the desired closed hyperplane.

From this follows immediately

Theorem 3 (Support theorem). *If x is not an interior point of a convex K which has interior points, then through x there is a closed hyperplane H such that K lies on one side of H.*

Theorem 4 (Eidelheit separation theorem). *Let K_1 and K_2 be convex sets in L such that K_1 has an interior point and K_2 contains no interior point of K_1. Then there is a closed hyperplane H separating K_1 from K_2; that is, there is a f in L^* such that* $\sup f(K_2) \leqq \inf f(K_1)$.

Proof. Let $K=K_1-K_2$; then K has interior points and 0 is not one of them. By Theorem 3 there is an f in L^*, $f \neq 0$, such that $f(x) \geqq 0$ in K. If $x_i \in K_i$, then $x_1-x_2 \in K$, so $f(x_1) \geqq f(x_2)$. Because R is a boundedly complete lattice, there is a c between $\inf\{f(x): x \in K_1\}$ and $\sup\{f(y): y \in K_2\}$. Set $H=\{x: f(x)=c\}$. $f(x)>c$ in the interior of K_1.

The support theorem can be improved in an LCS.

Theorem 5. *If K is a closed convex set in an* LCS *L and if x is not in K, then there is an f in L^* such that* $f(x)>\sup\{f(y): y \in K\}$.

Proof. Take a convex neighborhood K_1 of x which has no point in common with $K=K_2$ and use the separation theorem.

Corollary 2. *If E is a non-empty subset of an* LCS *L, then $K(E)$, the smallest closed convex set containing E, is the intersection of all the closed half-spaces containing E; that is,*

$$K(E)=\bigcap_{f \in L^*}\{x: f(x) \leqq \sup\{f(y): y \in E\}\}.$$

This corollary implies

Corollary 3. *In an* LCS, $(E^{\pi})_{\pi}=K(E \cup \{0\})$, *so if W is a wedge in L, then ${W^{\pi}}_{\pi}$ is the smallest closed wedge in L containing W, and if E is linear in L, then ${E^{\pi}}_{\pi}$ is the smallest closed linear subspace of L containing E.*

Corollary 4. *In a locally convex space every closed convex set is weakly closed.*

Corollary 4 can be rephrased in terms of dual spaces as

Corollary 5. *The closure of a convex set K in an* LCS *L is determined by the space L^*; no more information about the locally convex topology in L is needed.*

The proof of the separation theorem for convex sets shows that both Eidelheit and Mazur theorems follow easily from the following simple special case of Cor. 1.

Theorem 6 (Existence theorem). *If W is a wedge with non-empty interior and $W \neq L$, then there is a non-trivial monotone f in L^*; that is, there is a closed hyperplane H through zero such that W lies on one side of H.*

To derive the Eidelheit and Mazur theorems from this take $W = \bigcup_{\lambda \geqq 0} \lambda(K-E)$ in the Mazur theorem and $W = \bigcup_{\lambda \geqq 0} \lambda(K_1 - K_2)$ in the Eidelheit theorem. Then W has an interior point but 0 is not an interior point of W. Theorem 6 asserts that an $f \neq 0$ exists in L^* such that $f(x) \geqq 0$ if $x \in W$. Then $f(x) \geqq f(y)$ if $x \in K[K_1]$ and $y \in E[K_2]$. To complete the proof of the Mazur theorem requires only the observation that an f in $L^\#$ which is bounded on a flat set E is constant on E, and that an f in $L^\#$ which is of constant sign on an open set is continuous (Cor. 4,2).

(6) Theorem 6 is a simple consequence of the separation and support theorems.

(7) If W has an interior point and f is a non-zero element of $L^\#$ which is non-negative on W, then $f(x) > 0$ at every interior point x of W and $f \in L^*$.

(8) The Hahn-Banach theorem follows directly from Theorem 6. [Let $M = L \times R$, $E = \{(x, f_0(x)) : x \in L_0\}$. $K = \{(x,r) : r \geqq p(x)\}$, $W = K - E$. Then the core of K is $\{(x,r) : r > p(x)\}$ so $(0,1)$ is a core point of K and of W. $(0,0)$ is not a core point of W. Theorem 6 gives a non-trivial monotone F in $M^\#$. $H = F^{-1}(0)$ contains no core point of W, by (7). Set $f(x) = r$ if $F(x,r) = 0$ to get an extension f of f_0; $(x,r) \in H$ implies $f(x) = r \leqq p(x)$, so f is dominated by p.]

(9) An example shows that some assumption is needed in this series of related results. Let $L = \sum_{i < \infty} L_i$, where each $L_i = R$; define C in L as follows: $x = (\xi_i) \in C$ if the last non-zero ξ_i is positive. (Then $0 \in C$ by default.) Then (a) C is a cone in L. (b) There is no core point in C. (c) $C \cap -C = \{0\}$ and $C \cup -C = L$. (d) There is no non-zero element of $L^\#$ which vanishes on C. (e) Every hyperplane in L meets C. (C is a *ubiquitous* set in the terminology of Klee [1].)

(10) In any LCS with a wedge it is often convenient (see Chapter VI) to have the wedge closed. The cone of the preceding example is not even *lineally closed;* that is, a line in L need not meet C in a segment closed

in the line. For example, if $x=(0,1,0,0,\ldots)$ and $y=(-1,0,0,\ldots)$, then $tx+(1-t)y\in C$ if and only if $t>0$. For discussions of the operation of lineal closure see Klee [5] and Nikodym.

(11) If W is a wedge in L and W^π is the polar set in L^*, then (a) $W^\pi=\{f: f\in L^* \text{ and } f(x)\geqq 0 \text{ for all } x \text{ in } W\}$. (b) W^π is a w^*-closed wedge in L^*.

(12) Tukey showed that two closed convex disjoint sets in a reflexive Banach space can be separated by a closed hyperplane. Dieudonné [5] showed that this property fails in $l^1(\omega)$. Klee [2] shows how local compactness is needed in the separation theorem when no interior point is available.

Chapter II. Normed Linear Spaces

§ 1. Elementary Definitions and Properties

Each of the function spaces mentioned in the introduction of the preceding chapter has (with one exception, $A(D)$) a norm $\| \ \|$ [Definition I, 3, 1] which defines the topology of major interest in the space; a neighborhood basis of a point x is the family of sets $\{y: \|x-y\| \leqq \varepsilon\}$ where $\varepsilon > 0$.

Hereafter we shall use N for a normed space, that is, a linear space in which a norm is already assigned. If the normed space is *complete*[1] under the metric $\|x-y\|$, then the space will be called a *Banach space*, and will generally be denoted by B. U will generally stand for the *unit ball*, $\{x: \|x\| \leqq 1\}$, unless otherwise noted; the *unit sphere* is the set $\{x: \|x\| = 1\}$.

The properties we have discussed in linear topological spaces sometimes have simpler character in normed spaces.

(1) A set E in a normed space is bounded if and only if it lies in some ball.

(2) Call a linear operator T from one normed space N into another N' *bounded* if it is bounded on the unit ball U in N; define $\|T\|$, the *bound* or *norm* of T, to be $\sup\{\|Tx\|: \|x\| \leqq 1\}$. Then: (a) $\|Tx\| \leqq \|T\| \cdot \|x\|$ if $x \in N$. (b) $\|T\| = \sup\{\|Tx\|: \|x\| = 1\} = \sup\{\|Tx\|: \|x\| < 1\} = \sup\{\|Tx\| / \|x\|: x \neq 0\}$. (c) T is bounded if and only if it is continuous, and if and only if it is uniformly continuous. (d) If T' is a bounded linear operator from N' into N'', then $\|T'T\| \leqq \|T'\| \cdot \|T\|$. (e) If T is a bounded linear operator from N into B', and if B is the completion of N, then T has a unique continuous linear extension T' from B into B', and $\|T'\| = \|T\|$.

(3) Let $\mathfrak{L}(N, N')$ be the space of all continuous linear operators from N into N'. (a) $\mathfrak{L}(N, N')$ is a normed linear space. (b) $\mathfrak{L}(N, N')$ is a Banach space (that is, complete) if and only if N' is. (c) Hence N^* is a Banach space. (d) This norm determines the $\mathfrak{B}^\pi$ topology in N^*.

---

[1] That is, every Cauchy sequence in N has a limit in N.

(e) Every bounded set E in $\mathfrak{L}(N,N')$ is a uniformly equi-continuous set of functions on N into N'; that is, given $\varepsilon>0$ there is $\delta>0$ such that for all T in E and all x,y in N

$$\|Tx-Ty\|<\varepsilon \quad \text{if} \quad \|x-y\|<\delta.$$

From the Hahn-Banach theorem [I, § 3] we get various simple results.

(4) Let L be a linear subspace of N. (a) If $f\in L^*$, then there is an extension F of f such that $F\in N^*$ and $\|F\|=\|f\|$. (b) If x is at distance d from L, then there exists an f in N^* such that f vanishes on L, $\|f\|=1$, and $f(x)=d$. (c) If $x\in N$, then $\|x\|=\sup\{|f(x)|: \|f\|=1\}=\sup\{|f(x)|: \|f\|\leqq 1\}$. (d) If E is a subset of N such that linear combinations of elements of E are dense in N, then in N^* there is a total set E' of functionals such that E and E' have the same cardinal number of elements. The converse is false; by § 2 there is no countable subset of $m(\omega)$ which is dense, but there is a countable set in $m(\omega)^*$ which is total over $m(\omega)$; this set is determined by any countable dense subset of $l^1(\omega)$. (e) Every finite-dimensional subspace L' of N^* has a finite-dimensional "approximate norming subspace" L in N; more precisely, given $\varepsilon>0$, for each f in L' there is x of norm one in L such that $f(x)>(1-\varepsilon)\|f\|$. [Take a set $f_1,\dots,f_n$ which is $\varepsilon/2$-dense in the unit sphere of L'; the definition of $\|f\|$ gives for each $i\leqq n$ a unit vector x_i in N with $f_i(x_i)>(1-\varepsilon/2)$. Let L be the subspace of N spanned by the x_i, $i\leq n$.] (f) Hence, if L' is a separable subspace of N^*, it has a separable norm-determining subspace L in N; that is, $\|f_{|L}\|=\|f\|$ for all f in L'. [Take an increasing sequence of n-dimensional subspaces L'_n whose union is dense in L'; construct L_n in N corresponding to L'_n and $\varepsilon_n=1/n$ by (e); let L be the sum of the L_n.] (g) If E is a separable subspace of N, then (f) applied to $Q(L)$ in N^{**} gives in N^* a separable norming subspace L' for N. [This is stronger than (d) which only implies that there is a separable total subspace of N^*.]

Theorem 1. *The function Q, defined by $Qx(f)=f(x)$ for all f in N^*, is a linear isometry of N into N^{**}.*[2]

Proof. It is already known, since N is locally convex and by (3d) the norm topology of N^* is the $\mathfrak{B}^\pi$ topology of N^*, that Q is $1-1$ linear from N into N^{**}. To check the isometry property, for each x in N

$$\|Qx\|=\sup\{|Qx(f)|: \|f\|\leqq 1\}=\sup\{|f(x)|: \|f\|\leqq 1\}=\|x\|$$

by (4c).

---

[2] When $Q(N)=N^{**}$, N is called *reflexive*; completeness of N is necessary but not sufficient for this; see III, § 4.

Theorem 2. *If $(x_n, n\in\Delta)$ is a net in N which converges weakly to x, then $\|x\| \leqq \liminf_{n\in\Delta}\|x_n\|$; that is, the norm is a weakly lower-semicontinuous functional in N. Dually, if $w^*\text{-}\lim_{n\in\Delta} f_n = f$, then $\|f\| \leqq \liminf_{n\in\Delta}\|f_n\|$; that is, a conjugate norm is w^*-lower-semi-continuous.*

Proof. To prove the harder one, take $\varepsilon>0$ and find x in N such that $\|x\|=1$ and $f(x)>\|f\|-\varepsilon$. Then

$$\|f_n\| \geqq f_n(x) \to f(x) > \|f\| - \varepsilon,$$

so

$$\|f_n\| > \|f\| - \varepsilon \quad \text{when } n > n(\varepsilon).$$

Every linear subspace of a normed space, when provided with the norm induced by the large space, is also a normed space; a norm can be introduced in a factor space of a normed space, as follows:

Definition 1. If E is a flat, closed subset of a normed linear space N, let $\|E\| = \inf\{\|x\| : x\in E\}$.

Lemma 1. *If N_0 is a closed linear subspace of N, if $N_1 = N/N_0$ is normed by the definition above, and if T_0 is the natural mapping $T_0 x = x + N_0$, then* (a) *N_1 is a normed linear space,* (b) *T_0 is continuous and interior* (Def. 3, 2), *so the norm defines the proper factor-space topology,* (I, 4, (7)), (c) *$\|T_0\| \leqq 1$ and T_0^* is a linear isometry of N_1^* onto $N_0^\perp \cap N^*$,* (d) *N_1 is complete if N is.*

Proof. Part (a) requires only straightforward calculations with the properties of norm in N and inf in the real numbers. $\|T_0 x\| \leqq \|x\|$ by definition, so $\|T_0\| \leqq 1$. But if $\varepsilon>0$, and $\|T_0 x\| < \varepsilon$, then there is a y in $x+N_0$ such that $\|y\|<\varepsilon$, hence $T_0(U)$ contains the open ε-ball about $T_0 z$ if U contains the ε-ball about z. This proves that T_0 is continuous and interior, so the norm determines the topology in N/N_0 which is appropriate to that factor space. We have proved $\|T_0\| \leqq 1$ so $\|T_0^* F\| \leqq \|F\|$ if $F\in N_1^*$. Then for each F in N_1^*, if $\varepsilon>0$ is given, there exists X such that $F(X) > \|F\| - \varepsilon$ and $\|X\| \leqq 1$. Also there exists x in X such that

$$\|x\| < \|X\| + \varepsilon \leqq 1 + \varepsilon; \quad \text{then}$$
$$(T_0^* F)(x) = F(T_0 x) = F(X) > \|F\| - \varepsilon;$$

hence $\|T_0^* F\| > (\|F\| - \varepsilon)/(1+\varepsilon)$; let $\varepsilon\to 0$ to get $\|T_0^* F\| \geqq \|F\|$ if $F\in N_1^*$. This, with the earlier inequality, shows that T_0^* is an isometry of N_1^* into N^*. Theorem I, 2, 3 shows that $T_0^*(N_1^*) \subseteqq N_0^\perp$. It remains to show for each $f\in N_0^\perp \cap N^*$ that there is a F with $T_0^* F = f$. But f is constant on cosets of N_0, hence the definition $F(X) = f(x)$ if $x\in X$ gives an element of $N_1^\#$; but $\|F\| = \sup\{|F(X)| : \|X\| < 1\} = \sup\{|f(x)| : x\in X$

and $\|X\|<1\}$, but $\|X\|<1$ implies there is x in X with $\|x\|<1$ so $|f(x)|<\|f\|$ if $x\in X$ and $\|X\|<1$. Hence $\|F\|\leqq\|f\|$ and $F\in N_1^*$. Clearly $T_0^* F=f$, so (c) is proved.

For (d) suppose N complete and let $(X_n, n\in\omega)$ be a Cauchy sequence in N_1. Then there exists an increasing sequence $(M_i, i\in\omega)$ of integers such that $\|X_m-X_n\|<2^{-i}$ if $m,n\geqq M_i$. Take $Y_0=X_{M_1}$, and $Y_i=X_{M_{i+1}}-X_{M_i}$ if $i\geqq 1$; choose y_i in Y_i so that $\|y_i\|<\|Y_i\|+2^{-i}$. Then

$$\Big\|\sum_{p<i\leqq q} y_i\Big\|<2\sum_{p<i\leqq q} 2^{-i}$$

so $\sum_{i\in\omega} y_i$ is convergent in the complete space N. If x is its sum and $X=T_0 x$, then $\|X_{m_{i+1}}-X\|\leqq\Big\|\sum_{j\leqq i} y_j-x\Big\|$. Thus the Cauchy sequence (X_n) has a subsequence converging to X, so (X_n) converges to X. Hence N_1 is complete.

A dual result is

Lemma 2. *If* $\mathfrak{i}$ *is the identity mapping into* N *of a closed linear subspace* N_0 *of the normed linear space* N, *then* $\mathfrak{i}^*$ *is a linear, continuous, interior mapping of* N^* *onto* N_0^* *with kernel* $\mathfrak{i}^{*-1}(0)=N_0^\perp$. *Hence* $\mathfrak{i}^*$ *determines a linear isometry* $\mathfrak{i}'$ *between* N_0^* *and* $N^*/N_0^\perp$; $\mathfrak{i}'(f)=\mathfrak{i}^{*-1}(f)$.

Proof. Because $\mathfrak{i}$ is linear and continuous, $\mathfrak{i}^*$ is linear. If $F\in N^*$, then $\mathfrak{i}^* F=F_{|N_0}$ (= F with its domain of definition reduced to N_0) so $\mathfrak{i}^{*-1}(0)=N_0^\perp$ and $\|\mathfrak{i}^* F\|\leqq\|F\|$ and $\mathfrak{i}^*$ is continuous; the Hahn-Banach theorem implies (4a) that if $f\in N_0^*$, then there is F in N^* with $\mathfrak{i}^* F=f$, and $\|F\|=\|f\|$. Hence $\mathfrak{i}^*$ is interior and carries N^* onto N_0^*. Lemma 1 can be used to verify that $\mathfrak{i}'$ is an isometry as well as an isomorphism.

As in the preceding chapter, two normed spaces N_1 and N_2 are *isomorphic* if there is a linear, one-to-one function from one space into the other which preserves the topology. Looking ahead to VII, § 1 for justification of the terminology, two normed spaces shall be called *isometric* if there is a linear isometry of one onto the other[3].

Corollary 1. *If* E *is a closed linear subspace of* N, *then* E^* *is isometric to* $N^*/E^\perp$ *and* $(N/E)^*$ *is isometric to* $E^\perp$.

Theorem 3. *If* T *is linear and continuous from one normed space* N *into another,* N_1, *then:* (a) T^* *is linear from* N_1^* *into* N^*, *and* $\|T^*\|=\|T\|$. (b) *If* $T(N)$ *is dense in* N_1, *then* T^* *is* $1-1$ *into* N^*. (c) *If* N *and* N_1 *are complete, and* $T(N)=N_1$, *then* N_1 *is isomorphic to* $N/T^{-1}(0)$, *and* T^* *is an isomorphism of* N_1^* *with* $N^*\cap T^{-1}(0)^\perp$. (d) *If* T *is an isomorphism of* N *with a subspace* N_2 *of* N_1, *then* T^* *is a continuous, interior mapping*

[3] Banach uses "equivalent" for this concept; Bourbaki uses "isomorphic".

of N_1^* onto N^* which has kernel $T^{*-1}(0)=N_2^{\perp}\cap N_1^*$. In particular, the adjoint of an isomorphism onto is an isomorphism onto.

Proof. Parts (a) and (b) use only straightforward calculations. For part (c) let $N_0=T^{-1}(0)$, let T_0 be the natural function carrying N onto N/N_0, and write as in I, 2, (4), $T=T_1 T_0$; then T_0 is interior, and it is easily verified that $\|T_1\|=\|T\|$, so T_1 is continuous and one-to-one. The continuity of T_1^{-1} requires a category proof; this is given in a later section, Theorem 3, 4. Most of (d) is also formal; the interiority proof uses (c) or Theorem 3, 4.

Some of what we propose to do in § 3 can be carried out in linear metric spaces.

Definition 2. M is a *linear metric space* (LMS) if it is a linear space such that: (A) it has defined in it a distance function or metric satisfying the usual conditions [Kelley, p. 118] $(A_1)\, d(x,y)=0$ if and only if $x=y$, $(A_2)\, d(x,y)=d(y,x)$, and $(A_3)\, d(x,y)+d(y,z)\geqq d(x,z)$ for all x, y, z in M, and (B) distance and linear structure are related by $(B_1)\, d(x-z,y-z)=d(x,y)$ for all x, y, z in M; that is, the metric is invariant under translation by an element of the space. (B_2) If $a_n\to 0$, $d(a_n x,0)\to 0$ for each x in M, and (B_3) if $d(x_n,0)\to 0$ and a is real, then $d(a x_n,0)\to 0$.

(5) To prove from these axioms that M is an LTS goes in several steps. (a) (B_1) implies continuity of addition. (b) (B_1) and (B_2) imply that multiplication is continuous in the first variable; (c) (B_1) and (B_3) give continuity in the second variable. (d) A category argument on the sequence of real-valued functions $d(a x_n,0)$, $|a|\leqq 1$, gives continuity of multiplication at $(0,0)$. (See § 3 for category proofs.) (e) (B_1) and (d) give continuity of multiplication, the only missing condition. (f) Replacing d by a new metric d', defined by $d'(x,0)=\sup\{d(ax,0): |a|\leqq 1\}$ and $d'(x,y)=d'(x-y,0)$ gives a metric which is topologically and uniformly equivalent to the old but has the extra property that $d'(ax,0)$ is a nondecreasing function of $|a|$.

(6) Any invariant metric in an LTS L which yields the original topology of L also yields the original uniform structure of L. [Klee 7.]

(7) If S is an index set let R^S have the product space topology. (a) R^S is an LCS. (b) R^S is metrisable if and only if S is countable. (c) R^S is locally bounded if and only if S is finite.

(8) Let D be a domain in the complex plane and let $A(D)$ be the vector space of (complex-valued) functions analytic in D. For K compact in D define p_K in $A(D)$ by $p_K(x)=\sup\{|x(t)|: t\in K\}$. Then: (a) Each p_K is a pre-norm in $A(D)$; if K is not a finite set, p_K is a norm in $A(D)$. (b) If a neighborhood system of 0 is defined by choosing all the sets of the form $\{x: p_K(x)<\varepsilon\}$ where K runs over compact subsets of D and

$\varepsilon>0$, the resulting topology is first-denumerable, hence metrisable. (c) Every Cauchy sequence (in this topology) has a limit in $A(D)$. (See Köthe [1, 2] for references and a discussion of $A(D)^*$.)

(9) If $0<p<1$, (a) $l^p(S)$ (Def. 2.1) is locally bounded but is not locally convex unless S is finite; (b) if μ is a measure on a Borel family of sets, then the space of μ-measurable, pth power μ-integrable functions $L^p(\mu)$, is locally bounded; $L^p(\mu)$ is locally convex if and only if it is finite dimensional; (c) when μ is Lebesgue measure $L^p(\mu)^*=\{0\}$. (See Day [10].)

(10) If N is a complex-linear normed space the correspondence of I, 1, (7), 2, (6), and 4, (18) between $(N_{(r)})^*$ and $(N^*)_{(r)}$ is an isometry.

(11) Mazur's theorem shows that every point x of the boundary of a closed convex set K with interior points is a support point of K; that is, there is an f in L^* with $f(x)=\sup f(K)$. Bishop and Phelps [1, 2] show that *the support points of a bounded closed convex set K in a Banach space B are dense in the boundary of K*, and that *the functionals supporting K are dense in B**. The proof depends on the construction of *support cones* of the form $x+K(f,k)$ where, if K is the unit ball of B, for $k>0$ and f of norm one in B^*, $K(f,k)=\{x\colon \|x\|\le kf(x)\}$. Completeness of B, or at least of K, is required for this result.

§ 2. Examples of Normed Spaces; Constructions of New Spaces from Old

Definition 1. A function $\{x_s, s\in S\}$ on a set S with values in a linear topological space L is called *unconditionally summable to an element x of L* if $\lim_{\sigma\in\Sigma}\sum_{s\in\sigma} x_s=x$, where Σ is the system of finite subsets of S directed by $\supseteqq$. Then we write $x=\sum_{s\in S} x_s$, and say that x is *the sum of the unconditionally convergent series* $\sum_{s\in S} x_s$. (See also IV, § 1.)

(1) If $L=R$, then $\sum_{s\in S} x_s$ is unconditionally convergent to some $x\in R$ if and only if there is a K such that $\sum_{s\in\sigma}|x_s|\leqq K$ for all σ; then only countably many x_s are different from 0. [See Kelley, p. 77, G.]

If S is a set of indices, certain spaces of real-valued functions on S can be defined, some of them in terms of unconditional convergence.

$m(S)=$ set of all bounded real-valued functions x on S; $\|x\|=\sup\{|x(s)|\colon s\in S\}$.

$c_0(S)=$ closed linear subspace of all those x in $m(S)$ such that for each $\varepsilon>0$, $\{s\colon |x(s)|>\varepsilon\}$ is finite.

$l^p(S)$ = set of real-valued functions whose pth power is unconditionally summable on S, with norm defined by $\|x\| = \left(\sum_{s \in S} |x(s)|^p\right)^{1/p}$.

If S is a topological space, $C(S)$ is the linear subspace of continuous functions in $m(S)$.

If in S is given a Borel field (Boolean σ-algebra) of subsets on which a non-trivial measure μ is defined [see Halmos [1] or Munroe or any other book on measure and integration], two functions are *equivalent* if their difference is zero except on a set of μ-measure zero. Then $L^p(\mu)$ is the space of all equivalence classes of μ-measurable functions x whose pth powers are μ-integrable; $\|x\| = \left[\int_S |x(s)|^p \, d\mu(s)\right]^{1/p}$. $M(\mu)$ is the space of equivalence classes of bounded measurable functions; $\|x\| = \inf\{k\colon \mu\{s\colon |x(s)| > k\} = 0\}$.

All the spaces just described are complete normed linear spaces; for most of them the proof of completeness comes in Theorem 1, below.

(2) Recall that a metric space is called *separable* if it has a countable dense subset. (a) S is countable if and only if $l^p(S)$, $p \geqq 1$, and $c_0(S)$ are separable. (b) $m(S)$ is separable if and only if S is finite. (c) $C[0,1]$ is separable, as is (d) $L^p(\mu)$ when μ is Lebesgue measure, or whenever the Borel field of measurable sets is generated, up to sets of measure zero, by a countable collection of measurable sets.

(3) For each s in S let f_s be defined on $B = m(S)$ or $l^p(S)$ by $f_s(x) = x(s)$ for all x in B. (a) Each f_s is in B^* and $\|f_s\| = 1$. Finite linear combinations $f = \sum_{s \in \sigma} t_s f_s$ are also in B^* and have norms as follows: (b) In $m(S)^*$ or $c_0(S)^*$, $\|f\| = \sum_{s \in \sigma} |t_s|$. (c) If $p > 1$, let q satisfy $1/p + 1/q = 1$; then in $l^p(S)^*$ (by Hölder's inequality) $\|f\| = \left(\sum_{s \in \sigma} |t_s|^q\right)^{1/q}$. (d) In $l^1(S)^*$, $\|f\| = \sup\{|t_s|\colon s \in \sigma\}$.

From these calculations follows certain useful isometries, which help to clarify the structure of certain conjugate spaces; Banach gives the case $S = \omega$.

Theorem 1. *$l^1(S)$ is linearly isometric with $c_0(S)^*$ under the mapping $Ty = f$ if for every x in $c_0(S)$*

$$f(x) = \sum_{s \in S} y(s)\, x(s).$$

Under the same sort of mapping $m(S)$ is equivalent to $l^1(S)^*$, and, if $p > 1$ and $1/p + 1/q = 1$, $l^q(S)$ is isometric to $l^p(S)^*$.

Proof. For the first case, let σ be a finite subset of S and define P_σ, a projection of $c_0(S)$, by $P_\sigma x(s) = x(s)$ if $s \in \sigma$, $=0$ if $s \in \sigma$. Ordering the stack Σ by $\supseteqq$ we have $\lim_{\sigma \in \Sigma} \|P_\sigma x - x\| = 0$ for every x in $c_0(S)$.

If δ^s is that element of $c_0(S)$ which is 1 at s and 0 elsewhere, and if $f \in c_0(S)^*$, define Uf to be the y on S for which $y(s) = f(\delta^s)$. Then for each x and σ,

$$P_\sigma^* f(x) = f(P_\sigma x) = \sum_{s\in\sigma} f(P_{\{s\}} x) = \sum_{s\in\sigma} f(x(s)\delta^s) = \sum_{s\in\sigma} y(s)\, x(s).$$

Therefore

$$\|f\| \geqq \|P_\sigma^* f\| = \sum_{s\in\sigma} |y(s)|$$

by (3 b), so by (1), $\sum_{s\in S} y(s)$ exists. We see that U carries $c_0(S)^*$ into $l^1(S)$ and that $\|U\| \leqq 1$. On the other hand T maps $l^1(S)$ linearly into $c_0(S)^*$ and $\|T\| \leqq 1$, because if $\|x\|_{c_0} \leqq 1$, then

$$\begin{aligned} |Ty(x)| &= \lim_\sigma |f(P_\sigma x)| = \lim_\sigma \Big|\sum_{s\in\sigma} y(s)\, x(s)\Big| \\ &\leqq \lim_\sigma \sum_{s\in\sigma} |y(s)|\, |x(s)| \leqq \lim_\sigma \sum_{s\in\sigma} |y(s)| = \|y\|_{l^1}. \end{aligned}$$

Proofs for the other isometries are similar. We remark that these relations are sometimes stated less precisely as: l^1 is the conjugate space of c_0, and so on.

(4)[4] If $p > 1$ and $1/p + 1/q = 1$, then $L^q(\mu)$ is linearly isometric to $L^p(\mu)^*$.

(5)[5] If μ is a measure and if $\mathbf{M}(\mu)$ is the set of bounded real functions measurable on every part of S which is a countable union of sets of finite μ-measure, then $\mathbf{M}(\mu)$ is isometric to the space $L^1(\mu)^*$.

Note that these proofs when completed imply that all the spaces described in this section except $L^1(\mu)$, $c_0(S)$, and $C(S)$, are isometric to conjugate spaces; the last two examples are closed subspaces of $m(S)$, hence they are also complete.

(6) If T is the mapping of Theorem 1 of $c_0(S)^*$ onto $l^1(S)$ and if U is the similar isometry of $m(S)$ onto $l^1(S)^*$, then the mapping $T^* U$ is an isometry of $m(S)$ onto $c_0(S)^{**}$ which coincides on $c_0(S) \subseteqq m(S)$ with the natural mapping Q of $c_0(S)$ into $c_0(S)^{**}$.

(7) If $x \in m(S)$, then there exist $x_\sigma = P_\sigma x$ in $c_0(S)$ such that $\sum_{s\in S} x(s)\, y(s) = \lim_\sigma \sum_{s\in\sigma} x_\sigma(s)\, y(s)$ for every y in $l^1(S)$. (This is related to a special case of Theorem 4,3.)

---

[4] See Banach, p. 65, for one proof for Lebesgue measure on an interval. Or approximate by functions constant on the sets of a finite family of disjoint measurable sets of finite measure, use (3), and work back.

[5] See, for example, Hewitt and Ross, Chapter III, for the non-σ-finite case.

(8) For each F in $m(S)^*$ define a function $TF=\varphi$ on $\mathbf{A}(S)$, the family of all subsets of S, by $\varphi(E)=F(\chi_E)$ for each E in $\mathbf{A}(S)$, where χ_E is the characteristic function of E. (a) Such a φ is finitely additive and of bounded variation; that is, if $E_1\cap E_2=\varphi$, then $\varphi(E_1\cup E_2)=\varphi(E_1)+\varphi(E_2)$, and the *total variation of* φ *on* S, $V\varphi(S)=\sup\left\{\sum_{i\leqq n}|\varphi(E_i)|: E_i \text{ pairwise disjoint subsets of } S\right\}$, is finite. (b) $\mathbf{BV}(S)$, the set of all finitely additive functions of bounded variation on $\mathbf{A}(S)$, is a linear space under setwise addition and scalar multiplication: $(\varphi+\psi)(E)=\varphi(E)+\psi(E)$ and $(\lambda\varphi)(E)=\lambda(\varphi(E))$ for all E in $\mathbf{A}(S)$. (c) $\mathbf{BV}(S)$ is a complete normed linear space if $\|\varphi\|=V\varphi(S)$. (d) Each φ in $\mathbf{BV}(S)$ defines a function $F=\int\ldots d\varphi$ in $m(S)^*$ by setting $F\left(\sum_{i\leqq n} c_i\chi_{E^i}\right)=\sum_{i\leqq n} c_i\,\varphi(E_i)$. This defines F on a dense linear subset L of $m(S)$; F is defined on the rest of $m(S)$ by uniform continuity: If $x_n\in L$ and $\|x_n-x\|\to 0$, then $F(x)=\lim_n F(x_n)$. (e) T is a linear isometry of $m(S)^*$ and $\mathbf{BV}(S)$. (Hildebrandt [1], Fichtenholz and Kantorovich.)

(9) (Riesz representation theorem.) If $S=[0,1]$, by the Hahn-Banach theorem, each f in $C[0,1]^*$ has an extension F of the same norm in $m([0,1])^*$; hence f can be represented by a finitely additive φ. Let $g(t)=\varphi[0,t]$; then g is of bounded variation, is right-continuous if $t>0$, and the Stieltjes integral $\int_0^1 x\,dg$ gives the original value $f(x)$ for every x in $C[0,1]$. (See Banach, p. 60, for a proof, and for reference to F. Riesz's original paper.) The association of g with $\int_0^1\ldots dg$ is an isometry between $C[0,1]^*$ and the set $\mathbf{bv}\,[0,1]$ of functions 0 at 0, right-continuous except at 0, and of bounded variation on $[0,1]$.

From these examples we can now construct more normed spaces, which shall be called *substitution spaces* or *product spaces.*

Definition 2. Let S be an index set. A *full function space on X* is a Banach space of (real or complex) functions ξ on S such that for each ξ in X each function η for which $|\eta(s)|\leq|\xi(s)|$ for each s in S is again in X and $\|\eta\|\leq\|\xi\|$. If also for each s in S a normed space N_s is given, let $P_X N_s$, the *substitution space of the N_s in X*, be the space of all those functions x on S such that (i) x_s is in N_s for each s in S, and (ii) if ξ is defined by $\xi(s)=\|x_s\|$ for all s in S, then $\xi\in X$. For each x in $P_X N_s$, define $\|x\|=\|\xi\|_X$.

(10) With the above definitions $P_X N_s$ is a normed linear space; it is complete if and only if all N_s are complete.

(11) Denote $P_{l^p(S)}N_s$ by $P_p N_s$. (a) If T is defined from $P_1(B_s^*)$ to $(P_{c_0(S)}B_s)^*$ by $Ty=f$ if $f(x)=\sum_{s\in S} y_s(x_s)$ for all x in $P_{c_0(S)}B_s$ (remember

that each y_s is a linear functional), then T is an isometry between the spaces mentioned. (b) Similarly $P_{m(S)}(B_s^*)$ is linearly isometric with $(P_1 B_s)^*$ and (c) when $p>1$ and $1/p+1/q=1$, $P_q(B_s^*)$ is isometric with $(P_p B_s)^*$, so (d) if $p>1$ then, $P_p B_s$ is a reflexive Banach space if and only if every B_s is reflexive.

(12) $l^1(S)$ has an elementary but peculiar property: If it is a homomorphic image of a Banach space B, then it is isomorphic to the range of a bounded linear projection of B into itself. More precisely, if T is a bounded linear operator from B onto $l^1(S)$ and if $L=T^{-1}(0)$, then the interior mapping theorem, II, 3, 4, asserts that $l^1(S)$ is isomorphic with B/L. Then there exists a $K<\infty$ such that for each basis vector δ^s in $l^1(S)$ there is an x^s in $T^{-1}(\delta^s)$ such that $\|x^s\|<K$. Setting $U((t_s))=\sum_{s\in S} t_s x^s$, the series is absolutely convergent in B for each (t_s) in $l^1(S)$. Then $U(l_1(S))$ is a closed subspace M of B, U is an isomorphism of $l^1(S)$ with M, and $U\circ T$ is the projection of B on M along L.

Every $L(\mu)$ space is sequentially complete in the w-topology (see Th. VI, 4, 4). $l^1(S)$ has a stronger property, which will be approached through a result of Phillips [1].

Lemma 1. *Let $(\Xi_n, n\in\omega)$ be a w^*-convergent to 0 sequence of elements from $c_0(S)^{***}$; then $\lim_{n\in\omega}\|Q^*\Xi_n\|=0$.*

Proof. First use the representation theorems for the conjugate spaces of $c_0(S)$. Each Ξ_n can be represented by a finitely additive set function φ_n of bounded variation on S; then $Q^*\Xi_n$ corresponds to ψ_n, the *atomic part* of φ_n, defined for each $E\subseteq S$ by $\psi_n(E)=\sum_{s\in E}\varphi_n(\{s\})$. Hence $\|Q^*\Xi_n\|=$ total variation of $\psi_n=\sum_{s\in S}|\varphi_n(\{s\})|$. The lemma can now be restated in an equivalent form:

(13) *Let $(\varphi_n, n\in\omega)$ be a sequence of finitely additive set functions of uniformly bounded variation on S such that for each $E\subseteq S$, $\lim_{n\in\omega}\varphi_n(E)=0$, and let ψ_n be the atomic part of φ_n; then $\lim_{n\in\omega}\|\psi_n\|=\lim_{n\in\omega}\sum_{s\in S}|\varphi_n(\{s\})|=0$.*

Proof. If this does not tend to zero, then there exist $\varepsilon>0$, and a subsequence (φ_n') on which $\sum_{s\in S}|\varphi_n'(\{s\})|\geqq\varepsilon$ for all n. Then disjoint finite sets $\sigma_k\subseteq S$ and a subsequence (Θ_k) of (φ_n') can be chosen so that

$$\sum_{s\in\sigma_k}|\Theta_k(\{s\})|>\sum_{s\in S}|\Theta_k(\{s\})|-\frac{\varepsilon}{10}\,.$$

Each Θ_k is finitely additive and of bounded variation, so the total variation $V\Theta_k$ shares these properties. Choose $\tau_1=\sigma_1$ and $\rho_1=\Theta_1$; divide the sequence $\sigma_2,\ldots,\sigma_k,\ldots$ into more than $10\,V\rho_1(S)/\varepsilon$ disjoint infinite sequences of sets. Then there is one of these subsequences $\sigma_{11}, \dots \sigma_{1n}, \dots$ such that $V\rho_1\left(\bigcup_{j\in\omega}\sigma_{1j}\right) < \varepsilon/10$. Let $\tau_2 = \sigma_{11}$, let k_2 be the place of σ_{11} in the sequence (σ_k), and let $\rho_2 = \Theta_{k_2}$. By induction this process determines a subsequence (τ_i) of (σ_k) and a subsequence (ρ_i) of (Θ_k) such that $\rho_i(\tau_i) > \sum_{s\in S}|\rho_i(\{s\})| - \varepsilon/10$ and $V\rho_i\left(\bigcup_{j>i}\tau_j\right) < \varepsilon/10$.

Define x in $m(S)$ by $x(s) = 0$ if $s \notin \bigcup_{i\in\omega}\tau_i$, $x(s) = (-1)^i \operatorname{sign}\rho_i(\{s\})$ if $s \in \tau_i$. Then

$$\left|\int_S x\,d\rho_i - (-1)^i \sum_{s\in\tau_i}|\rho_i(\{s\})|\right| \leqq \left|\sum_{j<i}\int_{\tau_i} x\,d\rho_i\right| + \left|\int_{\bigcup_{j>i}\tau_j} x\,d\rho_i\right|$$

$$\leqq \sum_{j<i}\sum_{s\in\tau_j}|\rho_i(\{s\})| + V\rho_i\left(\bigcup_{j>i}\tau_i\right)$$

$$\leqq \sum_{s\notin\tau_i}|\rho_i(\{s\})| + V\rho_i\left(\bigcup_{j>i}\tau_i\right) < \frac{\varepsilon}{10} + \frac{\varepsilon}{10}.$$

Hence the sequence (Ξ_n) has a subsequence (Ξ_{n_i}) such that $\Xi_{n_i}(x) = \int_S x\,d\rho_i$ oscillates from above $7\varepsilon/10$ to below $-7\varepsilon/10$ and back. This contradicts the original assumptlon that (Ξ_n) is w^*-convergent to zero, so Phillips' lemma is proved.

Corollary 1. *Let T be a continuous linear operator from $m(S)$ into a normed space B, and let $b_s = T\delta_s$ (where δ_s, the Kronecker δ, is defined by $\delta_s(s') = 1$ or 0 according as $s' = s$ or $s' \neq s$). Then for each sequence (ξ_n) in B^* such that $w^*\text{-}\lim_{n\in\omega}\xi_n = 0$ it follows that $\lim_{n\in\omega}\sum_{s\in S}|\xi_n(b_s)| = 0$.*

This follows from Phillips' lemma and the simple verification that the sth coordinate of $Q^*T^*\xi_n$ is $\xi_n(b_s)$.

Corollary 2. *In $l^1(S)$ weak and norm convergence of a sequence to an element are equivalent; that is, $w\text{-}\lim_{n\in\omega} f_n = f$ implies that $\lim_{n\in\omega}\|f_n - f\| = 0$.*

Proof. Let Q_1 be the natural map from $l^1(S)$ to its second conjugate; we have that $Q_1(f_n - f)$ has w^*-limit 0, and that Q^*Q_1 is the identity, so $\|f_n - f\|$ tends to O by Phillips' result.

Corollary 3. *$l^1(S)$ is weakly sequentially complete.*

Proof. The existence of $\lim_{n\in\omega} X(f_n) = F(X)$ for every X in $l^1(S)^*$ implies by Cor. II, 3, 1 that the sequence (f_n) is bounded; this in turn implies that F is an element of $l^1(S)^{**}$. Then $0 = w^*\text{-}\lim_{n\in\omega}(Q_1 f_n - F)$, so $\lim_{n\in\omega}\|f_n - Q^*F\| = 0$. Hence Q^*F is the weak limit of the weak Cauchy sequence (f_n).

Corollary 4. *There exists a Banach space B, a closed linear subspace L of B, and a sequence* $(f_i, i\in\omega)$ *in* L^* *such that* $w^*\text{-}\lim_{i\in\omega} f_i=0$, *but no matter how an extension* g_i *of each* f_i *is chosen in* B^*, $w^*\text{-}\lim_{i\in\omega} g_i$ *does not exist.*

Proof. Let $B=c_0(\omega)^{**}$ and $L=Q(c_0(\omega))$. For each i define f_i in L^* by $f_i(Qx)=i$th coordinate of x for each x in $c_0(\omega)$. Then $w^*\text{-}\lim_{i\in\omega} f_i=0$, and $\|f_i\|=1$ for all i. To say that g_i is an extension of f_i in B^* is to say that $f_i=Q^* g_i$. Phillips' lemma applies to prove that $w^*\text{-}\lim_{i\in\omega} g_i$ cannot exist, as it would imply $\lim_{i\in\omega} \|f_i\|=0$.

(14) Grothendieck [5] discovered another unusual property of $m(S)^*$: Every w^*-convergent *sequence* in $m(S)^*$ is also w-convergent (to the same limit). See VI, 4, (4), for a sketch of the proof (which uses Cor. 1) and for some generalizations.

§ 3. Category Proofs

The importance of category proofs in this subject depends largely on Theorems 3 and 4 of this section; the wide applications are due to the well-known theorem that every complete metric space is of second category in itself. To begin with a simple example, we have

Theorem 1. *A subset E of a normed space N is weakly bounded (or weakly totally bounded) if and only if it is contained in some ball in N. If N is complete, dual results hold for weak* (total) boundedness.*

Proof. If E is w-totally bounded, it is w-bounded (I, 4, (11c)). Then for f in N^* the set $f(E)$ is a bounded set of real numbers so $\bigcup_{n\in\omega} nE^\pi=N^*$. Each nE^π is closed, so at least one of these sets contains some ball; by symmetry and convexity of all nE^π it may be taken to be a ball U with center 0 and radius $\varepsilon>0$. Then for each x in E, $|f(x)|\leqq n$ if $f\in U$, that is $|g(x)|\leqq n/\varepsilon$ if $\|g\|\leqq 1$. Hence $\|x\|\leqq n/\varepsilon$ if $x\in E$, so E lies in a ball of N.

Conversely, if E lies in some ball in N, and if φ_π is a w-neighborhood of 0 in N, let $N_0=\bigcap_{f\in\varphi} f^{-1}(0)$; then $N_1=N/N_0$ is finite-dimensional. By Lemma 1,1, $E_1=T_0(E)$ is contained in some ball in N_1 and $U=T_0(\varphi_\pi)$ contains some neighborhood U_0 of 0 in N_1. Hence the closure of E_1 is compact and can be covered by a finite number of translates $\bigcup_{i\leqq n}(X_i+U_0)$. Then $E\subseteq\bigcup_{i\leqq n}(x_i+\varphi_\pi)$, where $x_i\in X_i$.

This category argument works as well for the w^*-topology provided N is complete. For an example to show the need for some completeness hypothesis, take N to be the space of real sequences $\xi=(x_i, i\in\omega)$ with only finitely many non-zero terms and with $\|\xi\|=\sup\{|x_i|:i\in\omega\}$. Then setting $f_i(\xi)=i x_i$ defines a norm-unbounded set in N^*, for $\|f_i\|=i$. However, for any ξ, $\lim_i f_i(\xi)=0$ so the set is w^*-bounded.

Corollary 1. *If* $(x_n, n\in\omega)$ *is a sequence of elements of* N *such that* $\limsup_{n\in\omega}|f(x_n)|<\infty$ *for every* f *in* N^*, *then* $\{\|x_n\|:n\in\omega\}$ *is bounded. If* $(f_n, n\in\omega)$ *is a sequence in* B^* *such that* $\limsup_{n\in\omega}|f_n(x)|<\infty$ *for every* x *in* B, *a complete normed space, then* $\{\|f_n\|:n\in\omega\}$ *is bounded.*

(1) Let $\mathfrak{L}=\mathfrak{L}(N,N')$ be the space of all linear continuous operators from N into N'; the *weak [strong] operator topology* in $\mathfrak{L}$ is the topology imposed on this set as a linear subspace of $(N')^N$ in its product topology, when the weak [norm] topology is used in N'. (a) If $(T_n, n\in\Delta)$ is a net of elements of $\mathfrak{L}$, then $\lim_{n\in\Delta} T_n=T$ in the weak [strong] operator topology if and only if for every f in N'^* and x in N

$$\lim_{n\in\Delta} f(T_n x)=f(Tx) \qquad \Bigl[\lim_{n\in\Delta}\|T_n x-Tx\|=0\Bigr].$$

(b) If $N'=R$, then weak and strong operator topologies both collapse to the weak*-topology of N^*. (c) If N is complete, the proof of Theorem 1 shows that for subsets of $\mathfrak{L}$ weak boundedness is equivalent to strong boundedness is equivalent to norm (sometime called uniform) boundedness. (d) Hence if $(T_n, n\in\omega)$ is a sequence of elements of $\mathfrak{L}$ such that $\limsup_{n\in\omega}|f(T_n x)|<\infty$ for every f in N'^* and x in N, then $\sup\{\|T_n\|: n\in\omega\}<\infty$. If Δ is a directed system with a countable cofinal subsystem, then $\limsup_{n\in\Delta}|f(T_n x)|<\infty$ for every f in N'^* and x in N implies that $\limsup_{n\in\Delta}\|T_n\|<\infty$.

The converses involve no category.

Definition 1. A set $E\subseteqq N$ is said to *span* N or to be *fundamental in* N if N is the smallest closed linear subset of N containing E.

Lemma 1. *If* $(T_n, n\in\Delta)$ *is a net of linear operators from a normed* N *into a Banach space* B *and if* $\limsup_{n\in\Delta}\|T_n\|=K<\infty$, *then the set* $E[E_w]$ *of those* x *for which* $Tx=\lim_{n\in\Delta} T_n x$ *exists in the norm [the weak] topology of* B *is linear and closed; on* E_w, *which contains* E, T *is linear and* $\|T\|\leqq K$.

Proof. By the vector properties of limits, E, E_w, and T are all linear. If $(x_i, i \in \omega)$ is a sequence of points of N with norm limit x, and if (weak or norm) limit $\lim_n T_n x_i = y_i$, then by Theorem 1,2, $\|y_i - y_j\| \leqq \liminf_{n\in\Delta} \|T_n\| \, \|x_i - x_j\| \leqq K \, \|x_i - x_j\|$, so (y_i) is a norm Cauchy sequence in B; by completeness (y_i) has a limit y in B. A typical rectangle approximation using the uniformity supplied by K does the rest; for example, if all x_n are in E, then

$$\|T_n x - y\| \leqq \|T_n(x - x_i)\| + \|T_n x_i - y_i\| + \|y_i - y\|.$$

The first and last terms can be made small by taking i large; then the middle is made small by taking n far out in Δ.

By Theorem 1,2 if $x \in E_w$, then $\|Tx\| \leqq K \, \|x\|$, so $\|T\| \leqq K$ when T is regarded as an element of $\mathfrak{L}(E_w, B)$.

Theorem 2. *Let B and B' be Banach spaces, let Δ be a directed system with a countable cofinal subset, and let $(T_n, n \in \Delta)$ be a net of linear continuous operators from B into B'. Then there is a linear continuous T from B into B' with* w-$\lim_n T_n x = Tx$ *for all x in B if and only if* (a) $\limsup_{n\in\Delta} \|T_n\| < \infty$ *and* (b) *there is a set E which spans B such that* w-$\lim_{n\in\Delta} T_n x$ *exists for every x in E.*

This follows from (1d) and Lemma 1.

The major results of this section are Banach's interior mapping theorem and the closed graph theorem, below.

Definition 2. Let T be a function from one Hausdorff topological space L into another L'. Then T is *continuous* [**cn**] if $T^{-1}(U)$ is open in L whenever U is open in L'; T is *interior* [**in**] if $T(U)$ is open in L' whenever U is open in L; T is *closed* [**cg**] whenever the graph $G(T) = \{(x, Tx) : x \in L\}$ is closed in $L \times L'$.

(2) (a) If T is **cn**, then T is **cg**. (b) If T is **in** and one-to-one, then T is **cg**. (c) In general, T **in** need not imply T **cg** nor **cn**; T **cn** need not imply T **in**; T **cg** need not imply T **cn** or **in**.

Theorem 3 (Closed graph theorem). *Let M and M' be linear metric spaces such that M is of second category in itself and M' is complete. If T is a closed linear operator defined on all of M into M', then T is continuous.*

Proof. In M' let S_ε be a closed ball about 0 with radius $\varepsilon > 0$. Then a category argument in M using the sequence of sets $(n T^{-1}(S_\varepsilon))$ shows that there is a $\delta = \delta(\varepsilon)$ with $0 < \delta < \varepsilon$ and $V_{\delta(\varepsilon)}$, the closed ball in M about 0 with radius $\delta(\varepsilon)$, contained in the closure of $A_\varepsilon = V_{\delta(\varepsilon)} \cap T^{-1}(S_\varepsilon)$.

Then take x in $V_{\delta(\varepsilon/2)}$. Approximate x by x_1 in $A_{\varepsilon/2}$ so closely that $x - x_1 \in V_{\delta(\varepsilon/4)}$. If $x_1, \ldots, x_i$ have been chosen so that $x - x_1 - \cdots - x_i \in V_{\delta(\varepsilon_i)}$, where $\varepsilon_i = \varepsilon/2^{i+1}$, choose x_{i+1} in A_{ε_i} so that $x - x_1 - \cdots - x_i - x_{i+1} \in V_{\delta(\varepsilon_{i+1})}$. Then $d(Tx_i, 0) \leqq \varepsilon/2^{i+1}$, so the Cauchy sequence $\left(\sum_{i \leqq n} Tx_i\right)$ converges to a y in M' such that $d(y,0) < \sum_{i\in\omega} \varepsilon/2^i = \varepsilon$. Also $d\left(\sum_{i\leqq n} x_i, x\right) \leqq \varepsilon 2^{-n-1}$, so $\sum_{i\in\omega} x_i$ converges to x. By hypotheses $y = Tx$, so $x \in T^{-1}(S_\varepsilon)$ if $x \in V_{\delta(\varepsilon/2)}$; this proves that T is continuous at 0 and this makes T continuous everywhere.

Theorem 4 (Banach's interior mapping theorem). *Let M and M' be linear metric spaces such that M is complete and M' is of second category in itself. Then every continuous linear T from M onto M' is interior, so M' is also complete if such a T exists.*

We shall prove this from Theorem 3 by certain formal juggling.

Definition 3. If L and L' are LTSs, say that L **cgn** L' whenever every closed linear operator from L into L' is continuous; say that L **cni** L' whenever every continuous linear operator from L onto L' is interior. If $\mathfrak{F}$ and $\mathfrak{F}'$ are families of LTSs, say that $\mathfrak{F}$ **cgn** $\mathfrak{F}'$ [$\mathfrak{F}$ **cni** $\mathfrak{F}'$] whenever for every L in $\mathfrak{F}$ and L' in $\mathfrak{F}'$ L **cgn** L' [L **cni** L'].

(3) If L, L_1, and L' are LTSs, if T_0 is a continuous, interior, linear function from L onto L_1, if T_1 is a linear function from L_1 into L', and if $T = T_1 T_0$, then T and T_1 have precisely the same ones of the following properties: (a) closed, (b) continuous, (c) interior, (d) onto L'.

(4) Let T be linear from one LTS L into another L', then (a) if T is closed, $T^{-1}(0)$ is also closed. (b) Whenever $L_0 = T^{-1}(0)$ is closed and $L_1 = L/L_0$, then T, factored as $T_1 T_0$, has the same of the properties of (3) as has T_1. (c) Hence T is closed if either (i) T is continuous or (ii) $T^{-1}(0)$ is closed and T is interior. (d) If L_0 is closed in L and L **cgn** [**cni**] L', then L/L_0 **cgn** [**cni**] L'.

(5) Suppose that $\mathfrak{F}$ and $\mathfrak{F}'$ are families of LTSs such that for each L in $\mathfrak{F}$ [in $\mathfrak{F}'$] and each closed linear subspace L_0 of L, the space L/L_0 is in $\mathfrak{F}$ [in $\mathfrak{F}'$]. Then (a) $\mathfrak{F}$ **cgn** $\mathfrak{F}'$ if and only if every one-to-one linear closed function from an L of $\mathfrak{F}$ into an L' of $\mathfrak{F}'$ is continuous. (b) $\mathfrak{F}'$ **cni** $\mathfrak{F}$ if and only if every continuous one-to-one linear map of an L' in $\mathfrak{F}'$ onto an L in $\mathfrak{F}$ is interior. (c) $\mathfrak{F}$ **cgn** $\mathfrak{F}'$ implies $\mathfrak{F}'$ **cni** $\mathfrak{F}$.

(6) If M is a metric linear space and if M_0 is a closed linear subspace, and if M_1 is the factor linear topological space M/M_0, then adaptations of the proofs of Lemma 1,1 prove: (a) the function $d(X,Y) = \inf\{d(x,y)\colon x \in X \text{ and } y \in Y\}$ makes the factor space an LMS; (b) this metric determines the factor space topology and uniform structure in the LTS M_1; (c) if M is complete, so is M_1; (d) if M is of second category in itself, so is M_1.

The proof of Theorem 4 follows from these. (6, c and d) assert that $\mathfrak{F}'$, the class of complete linear metric spaces, and $\mathfrak{F}$, the class of second-category-in-themselves linear metric spaces, satisfy the hypotheses of (5). Theorem 3 asserts that then $\mathfrak{F}$ **cgn** $\mathfrak{F}'$, so (5c) asserts that $\mathfrak{F}'$ **cni** $\mathfrak{F}$. This is Theorem 4.

For other work in this general field see Theorem III, 1, 2 and Kelley, [1, Chap. 6, P, Q, R], Ptak [1], Pettis [1], and, for other references, Dieudonné [1]. See also Husain and Robertson–Robertson.

(7) Other consequences of (5) and the continuity assertions of I, § 4, Cor. 4 (c) and (14b) are: (a) Every continuous linear function from an LTS onto a finite-dimensional LTS is interior. (b) Every continuous linear function from an LCS onto a space with the convex core topology is interior.

(8) If M is a complete linear metric space with distance function d and if a new and always smaller [or always larger] distance function d' is introduced under which M becomes a complete linear metric space M', the identity operator is an isomorphism of M with M'.

(9) The theorem of Löwig that every separable, infinite-dimensional, Banach space has a vector basis of the cardinal number of the continuum, and the fact that two such spaces may fail to be isomorphic (say $l^1(\omega)$ and $l^2(\omega)$), shows that some relation between the metrics is needed in (8).

(10) A *keg*[6] in an LCS L is a closed, convex, symmetric (or discoid) set containing 0 as a core point. An LCS is called *kegly* if every keg in L is a neighborhood of 0 in L. (a) A category argument shows that every Banach space, even complete LCM, is kegly.

(11) Topological Products of Banach spaces. In this book on normed spaces it is appropriate to consider results on locally convex spaces if either (a) the general case has the proof of the special normed case (like the Kreĭn-Mil'man theorem), or (b) the general case is easily proved from the normed case by the standard embedding into a product of Banach spaces which will be given in (f) (like (12), below, and Lemma III, 2, 1).

Let S be an index set and for each s in S let B_s be a Banach space; let M be the topological product $\prod_{a\in S} B_s$; that is, convergence in M is coordinatewise convergence. Then (a) *A set E is bounded in M if and only if each projection $P_s E$ is bounded in B_s.* (b) If T_s is the natural embedding of B_s into $M([T_s x](s')=x$ if $s=s'$, $=0$ if $s\neq s')$, then each T_s is continuous and linear, and the definition $f_s=f\circ T_s$ for each s determines as in I, 5, (9) an isomorphism V between M^* and $\prod_{s\in S} B_s^*$. (Note that this

[6] Bourbaki uses *tonneau* and *tonnelé*, now loosely translated as *barrel* and *barrelled.*

includes the information that if $f \in M^*$, then there is a finite set $\sigma \subset S$ such that $f(x)=0$ if $x(s)=0$ for all s in σ.) (c) Hence *the weak topology in M is the product topology of the weak topologies in the spaces B_s*. (d) By I, 2, (2b) there is a similar natural isomorphism between $M^{*\#}$ and $\prod_{s \in S}(B_s^{*\#})$ which carries $(M^*, \mathfrak{B}^\pi)^*$ onto $\prod_{s \in S} B_s^{**}$. (e) *M is complete;* that is, every Cauchy net has a limit. (f) If L is an LCS choose a suitable index set S and a neighborhood base $\{U_s: s \in S\}$ of 0 in L so that each U_s is convex and symmetric. Let p_s be the Minkowski functional of U_s; then p_s is a pre-norm in L which determines a norm in $N_s = L/p_s^{-1}(0)$ (but remember this may not determine the factor-space topology in this linear space), and in the completion B_s of N_s. Let $M = \prod_{s \in S} B_s$, and define the natural mapping $W = (W_s, s \in S)$ of L into M by: For each x in L, $W_s(x) = x + p_s^{-1}(0)$, the coset in N_s in which x lies. Then *W is an algebraic and topological isomorphism of L into M.* (g) *Hence L is complete if and only if $W(L)$ is closed in M.*

(12) (a) Mackey [2] showed by Theorem 1 and (11c) that in *every* LCS *L a set E is bounded if and only if it is weakly bounded.* (b) Another formulation of this is: *If two locally convex topologies in a linear space determine the same set of continuous linear functionals, then they determine the same bounded sets;* or (c) *the bounded sets in an* LCS *L depend only on L^*.* (d) The converse is not true.

§ 4. Geometry and Approximation

As we saw in §§ 3 and 6 of I, the Hahn-Banach theorem holds in a normed N to give separation and support theorems for convex bodies. As one application note

Theorem 1. *If E is a flat, closed, proper subset of N, then there is a closed hyperplane $H \supseteq E$ such that $\|H\| = \|E\|$.*

Proof. E does not meet the interior of the ball about 0 of radius $\|E\|$; by the Mazur theorem there is an H containing E, so $\|H\| \leqq \|E\|$, but H does not meet the interior of that ball, so $\|H\| = \|E\|$.

(1) When H is given as $f^{-1}(c)$, $f \in N^*$, then $\|H\| = |c|/\|f\|$.

For existence of a continuous linear extension of a function on a subset we have an analogue of Lemma I, 2, 1.

Theorem 2. *A real-valued function f defined on a set $X \subseteq N$ has a linear extension F of norm $\leqq M$ if and only if $\left|\sum_{i \leqq n} t_i f(x_i)\right| \leqq M \left\|\sum_{i \leqq n} t_i x_i\right\|$ for all choices of n in ω, scalars $t_1, \ldots, t_n$, and elements $x_1, \ldots, x_n$ of X.*

Proof. By Lemma I, 2, 1, f has a linear extension g defined on the linear set L of all linear combinations of points of X. By the hypothesis $|g(y)| \leqq M\|y\|$ if $y \in L$; by the Hahn-Banach theorem, with $M\|x\|$ for $p(x)$, g has an extension F with $\|F\| \leqq M$.

For a finite set X this yields

Corollary 1. *If $x_1, \ldots, x_n \in N$ and $c_1, \ldots, c_n \in R$, then there exists f in N^* with $f(x_i) = c_i$ and $\|f\| \leqq M$ if and only if $\left|\sum_{i \leqq n} t_i c_i\right| \leqq M \left\|\sum_{i \leqq n} t_i x_i\right\|$ for all choices of $t_1, \ldots, t_n$ in R.*

This corollary has an approximate dual, although, unless N is reflexive, the theorem has not. (See III, 4, (J).) A special case of this was proved by Helly [1].

Theorem 3. *If $f_1, \ldots, f_n \in N^*$ and if $M, c_1, \ldots, c_n \in R$, then for each $\varepsilon > 0$ there is an x in N with $f_i(x) = c_i$, $i = 1, \ldots, n$, and $\|x\| < M + \varepsilon$ if and only if $\left|\sum_{i \leqq n} t_i c_i\right| \leqq M \left\|\sum_{i \leqq n} t_i f_i\right\|$ whenever $t_1, \ldots, t_n \in R$. Rephrased geometrically, with $H_i = \{x : f_i(x) = c_i\}$, $E = \bigcap_{i \leqq n} H_i$, and*

$$M_0 = \sup \left\{ \frac{\left|\sum_{i \leqq n} t_i c_i\right|}{\left\|\sum_{i \leqq n} t_i f_i\right\|} : t_1, \ldots, t_n \in R \text{ and } \left\|\sum_{i \leqq n} t_i f_i\right\| \neq 0 \right\},$$

this says that $\|E\| = M_0$.

Proof. The first condition says that $\|E\| \leqq M$, the second that $M_0 \leqq M$, so equivalence of the two conditions on M is equivalent to $\|E\| = M_0$.

If the first condition holds and if $\varepsilon > 0$, take a corresponding x in E; then

$$\left|\sum_{i \leqq n} t_i c_i\right| = \left|\sum_{i \leqq n} t_i f_i(x)\right| \leqq \|x\| \cdot \left\|\sum_{i \leqq n} t_i f_i\right\| \leqq (M + \varepsilon) \left\|\sum_{i \leqq n} t_i f_i\right\|.$$

Let $\varepsilon \to 0$ to get $\left|\sum_{i \leqq n} t_i c_i\right| \leqq M \left\|\sum_{i \leqq n} t_i f_i\right\|$, or $M_0 \leqq M$.

Conversely, if M satisfies the second condition, either all $f_i = 0$ so that all $c_i = 0$ and $x = 0$ is in E, or else not all f_i are 0. Then there is a maximal linearly independent set of them which, with suitable renumbering, is $f_1, \ldots, f_r$, where r is an integer $\leqq n$. Any remaining f_i depend on these. By Cor. I, 2,2, $E' = \bigcap_{i \leqq r} H_i$, is not empty, and $E' = E$ because $E' \subseteqq H_i$ for $r < i \leqq n$. As an intersection of hyperplanes, E is flat; by Theorem 1 there is a hyperplane H in N such that $H \supseteqq E$ and $\|H\| = \|E\|$. But if $H = \{x : f(x) = c\}$, then by Corollary I, 2,4, $H \supseteqq E$

implies that there exist t_i in R such that $f=\sum_{i\leqq n} t_i f_i$ and $c=\sum_{i\leqq n} t_i c_i$. By (1)

$$\|E\|=\|H\|=\frac{|c|}{\|f\|}=\frac{\left|\sum_{i\leqq n} t_i c_i\right|}{\left\|\sum_{i\leqq n} t_i f_i\right\|}\leqq M_0\leqq M.$$

Corollary 2. *If $X\in N^{**}$, if $f_1,\ldots,f_n\in N^*$, and if $E=\bigcap_{i\leqq n} H_i$, where $H_i=\{x:f_i(x)=X(f_i)\}$, then $\|E\|\leqq\|X\|$.*

Proof. $\left|\sum_{i\leqq n} t_i X(f_i)\right|\leqq\|X\|\,\left\|\sum_{i\leqq n} t_i f_i\right\|$, so $M_0\leqq\|X\|$.

Corollary 3. *If $X\in N^{**}\setminus Q(N)$, if $H'=\{f:f\in N^*$ and $X(f)=c\}$, and if $M>|c|/\|X+Q(N)\|$, then the point 0 of N^* is in the weak*-closure of the part of H' in S_M, the ball in N^* about 0 of radius M.*

Proof. If φ is the set $\{x_i: i\leqq n\}$, then φ^π is a w^*-neighborhood of 0 in N^*; it contains the intersection E' of the hyperplanes H' and $H_i'=\{f:f(x_i)=0\}$. Cor. 2, lifted one space and applied to $X, Qx_1,\ldots,Qx_n$, shows that $\|E'\|<M$, so E' contains a point in S_M.

(2) When $X_1,\ldots,X_n\in N^{**}$, $c_1,\ldots,c_n\in R$, $H_i'=\{f: X_i(f)=c_i\}$, and $\mathfrak{X}_i=X_i+Q(N)\in N^{**}/Q(N)$, then $0\in w^*$-closure of $\bigcap_{i\leqq n} H_i'\cap S_M$ if and only if $M>\sup\left\{\left|\sum_{i\leqq n} t_i c_i\right|/\left\|\sum_{i\leqq n} i_i\mathfrak{X}_i\right\|: t_i \text{ real}\right\}$.

(3) Jameson gives estimates for the shape and size of the w^*-closure of $X_\perp\cap U$ which are more precise than Cor. 3 will yield.

(4) Lindenstrauss and Rosenthal show that the finite-dimensional subspaces of B^{**} are all approximated by finite-dimensional subspaces of B, in the following sense: For each finite dimensional subspace F of B^{**} and each $\varepsilon>0$ there is a subspace F_ε of B and an isomorphism T_ε of F on F_ε such that $\|T_\varepsilon\|\,\|T_\varepsilon^{-1}\|<1+\varepsilon$, and for each X in $F\cap Q(B)$, $T_\varepsilon X=Q^{-1}(X)$.

§ 5. Comparison of Topologies in a Normed Space

We have always available in N and in its conjugate spaces the norm and the weak topologies; in every conjugate space the w^*-topology is also present.

Lemma 1. *Norm and weak (and weak*) topologies agree in N (in N^*) if and only if N is finite-dimensional.*

Proof. If N is finite-dimensional, then all topologies agree. If norm and weak topologies agree, then the unit ball contains some φ_π

$=\{x: |f(x)| \leqq 1$ for all f in $\varphi\}$. But $\varphi_\pi \supseteqq \varphi_\perp$, and the deficiency of $\varphi_\perp$ in N is the number of linearly independent elements of φ (Cor. I, 2,3). If the dimension of N exceeds this number, then there is a non-zero element x of $\varphi_\perp$; then $tx \in \varphi_\perp$ for all t in R, but this is impossible because $\|tx\| = |t| \cdot \|x\| \leqq 1$.

Corollary 1. *In an infinite-dimensional normed space N, no w- or w*-neighborhood is bounded.*

Theorem 1 (Riesz [2]). *A normed space is finite-dimensional if and only if every bounded closed set is compact.*

Proof. This is a well-known property of Euclidean n-spaces; by I, 4, (11) it holds for all finite-dimensional spaces. If N is infinite-dimensional, take $\|x_1\|=1$. If $x_1, \dots, x_n$ have been chosen, let L_n be the linear hull of this set of x_i; then $L_n \neq N$ so there is an $f_{n+1} \neq 0$ such that $f_{n+1}(x_i)=0$ if $i \leqq n$. Take x_{n+1} so that $\|x_{n+1}\| \leqq 1$ and $f_{n+1}(x_{n+1}) \geqq \|f_{n+1}\|/2$. Then if $m>n$, $\|x_m - x_n\| \geqq f_m(x_m - x_n)/\|f_m\| \geqq 1/2$. Hence the unit ball in N is not compact.

As the example of N^* with its w^*-topology shows, there are many infinite-dimensional LCS's where *all bounded closed sets are compact.*

Even though weak and norm closure generally disagree in N, Cor. I, 6,5 shows that they agree for convex sets; this gives

Corollary 2 (Wehausen). *An infinite-dimensional normed space is the union of countably many sets, the integer multiples of the unit ball, which in the weak topology are closed and nowhere dense.*

Definition 1. Let $(x_n, n \in \Delta)$ be a net of elements of a linear space L; a net $(y_m, m \in \Delta')$ is called a *net of averages far out in* (x_n) if for each n in Δ there is an m_n in Δ' such that whenever $m > m_n$ then y_m is a convex combination, $y_m = \sum_{j \leq q} t_j x_{n_j}$, where all the $n_j \geq n$ and all $t_j \geq 0$ and $\sum_{j \leq q} t_j = 1$.

(1) If L is an LCS, if $x = \lim_{n \in \Delta} x_n$, and if $(y_m, m \in \Delta')$ is a net of averages of elements far out in (x_n), then $x = \lim_{m \in \Delta'} y_m$.

Theorem 2. *If L is an* LCS, *and if $\mathscr{U}$ is a convex neighborhood base at* 0, *and if x is the weak limit of a net $(x_n, n \in \Delta)$, then there is a net $(y_m, m \in \Delta')$ of averages far out in (x_n) such that x is the limit of y_m in the original topology of L. Δ' can be chosen to be*[7] $\Delta \times (\mathscr{U}, \subseteqq)$.

[7] The order relation to be used in a product of directed systems is the northeast ordering; that is, in this case $(n, U) \geq (n', U')$ means that $n \geq n'$ and $U \subseteq U'$.

Proof. Let E_n be the convex hull of $\{x_i : i \geq n\}$ and let F_n be the closure in L of E_n. By Cor. 1.6.4 every F_n is w-closed, so $x \in F_n$. Take $\Delta' = \Delta \times \mathscr{U}$ and for $m = (m, U)$ take y_m to be a point of $E_n \cap (x + U)$.

Norm-closed convex sets in a conjugate space need not be w^*-closed; a hyperplane $Z = \{f : X(f) = 0\}$ is w^*-closed if and only if $X \in Q(N)$.

Theorem 3 [Banach, Alaoglu]. *A norm-closed ball in N^* is w^*-compact, hence it is w^*-closed.*

Proof. The unit ball of N^* is U^π, the polar set of a neighborhood of 0 (the unit ball U of N); by Lemma I, 5, 3 U^π is w^*-compact. By w^*-continuity of vector operations, every norm-closed ball is w^*-compact.

Corollary 3. *The w^*-closure of a norm-bounded set in N^* is norm-bounded and w^*-compact.*

Corollary 4. *A convex set E in N^* is w^*-compact if and only if it is norm-bounded and w^*-closed.*

Proof. If E is w^*-closed and norm-bounded, Cor. 3 asserts that it is w^*-compact. If E is w^*-compact in N^*, then it is w^*-closed and w^*-bounded in N^*; if it is not also n-bounded, let B be the completion of N and apply Theorem 3, 1 to get a b in B and a sequence $(f_i, i \in \omega)$ in E such that the real sequence $(f_i(b), i \in \omega)$ is an unbounded increasing sequence. Find a sequence $(c_i, i \in \omega)$ such that all $c_i \geqq 0$, $\sum_{i \in \omega} c_i = 1$, and $\sum_{i \in \omega} c_i f_i(b)$ diverges. Because E is convex, $f = \sum_{i \in \omega} c_i f_i$ is in the w^*-closure of E in $N^\#$, but f is not in N^*, because it is unbounded in every neighborhood of b. Hence E is not w^*-closed in $N^\#$, and cannot be w^*-compact in N^*.

Theorem 4. *If U is the unit ball in a normed space, then $Q(U)$ is w^*-dense in $(U^\pi)^\pi$, the unit ball of N^{**}.*

This is an immediate consequence of Cor. 4, 2.

Corollary 5. *If T is a continuous linear operator on N into N' and if $U = \{x : \|x\| \leqq 1\}$, then $Q'(T(U))$ is a w^*-dense subset of $T^{**}(U^{\pi\pi})$ and $Q' \circ T = T^{**} \circ Q$.*

Proof. If $X \in U^{\pi\pi}$, take a net $(x_n, n \in \Delta)$ in U such that $X = w^*\text{-}\lim_{n \in \Delta} Q x_n$. Then for each f' in N'^*

$$T^{**} X(f') = X(T^* f') = \lim_{n \in \Delta} T^* f'(x_n) = \lim_{n \in \Delta} f'(T x_n) = \lim_{n \in \Delta} Q' T x_n(f').$$

Hence $T^{**} X = w^*\text{-}\lim_{n \in \Delta} Q' T x_n$.

Definition 2. The *bounded weak* [*bounded weak**] *topology* of N [of N^*] is the topology in which a set E is closed if and only if $E \cap A$ is w-[w^*-] closed in A whenever A is a norm-bounded subset of N [of N^*]; the corresponding closure of E will be denoted by $bw(E)$ [by $bw^*(E)$].

(2) (a) Knowledge of the w- or w^*-topology for bounded sets is not enough to determine it for all sets in N. For example, if $(L_n, n\in\omega)$ is an increasing sequence of n-dimensional subspaces of N, if E_n is the set of points of norm n in L_n, if A_n is a finite subset of E_n which contains a point within $1/n$ of each point of E_n, and if $A=\bigcup_{n\in\omega} A_n$, then 0 is in the w-closure of A, while every bounded part of A is finite, hence closed.

(b) bw-closedness and w-quasicompletness (see Defn. III, 1,1) can be defined in one step from convergence of bounded nets but bw-closure of a set in B cannot, because adding to E the limits of bounded w-convergent nets does not, usually, give an idempotent closure operation. (Sequential closure in R^S gives a reasonable analogy.) Wheeler [1] gives an example of a set E in c_0 which is bw-closed but at the same time is a proper subset of $Q^{-1}(bw^*(Q(E)))$. Let E be the set of all sequences x_{nk} where $x_{nk}(i)=0$ if $i>n$, $=1/k$ if $i<n$, and $=k$ if $i=n$. Then each bounded subset of E is w-closed (but not w-complete). In $m(=c_0^{**})$ for each k in ω there is a bounded net (even sequence) $((x_{nk}, n\in\omega)$ of elements of norm k with w^*-limit e/k, where $e=(1,1,\ldots,1,\ldots)$. Then there is a bounded net $(e/k, k\in\omega)$ with w^*-limit 0. Hence closing the second time gave points not in the first closure; also $0\notin Q(E)$. Hence the bw-topology of c_0 is not the image under Q^{-1} of the bw^*-topology relativized to $Q(c_0)$ but has more closed sets. (If E is *bounded*, of course this does not happen; that is, $Q^{-1}(bw^*(Q(E)))=Q^{-1}(w^*(Q(E)))=w(E)$.) (c) Nevertheless, iteration, usually transfinitely often, of adding the limits of bounded, w-convergent nets does give $bw(E)$, the smallest bw-closed set containing E. For a subset E of a [conjugate] normed space $N[N^*]$, the following conditions are equivalent: (i) E is bw-closed in N [bw^*-closed in N^*]. (ii) $E\cap V$ is w-[w^*-] closed for every closed ball V in N [in N^*]. (iii) E contains the bw- [bw^*-] limit of every bw- [bw^*-] convergent net in E. (iv) E contains the w- [w^*-] limit of every bounded w- [w^*-] convergent net in E. (d) If an LCS L is bw-complete, then it is w-quasi-complete; that is, every bounded w-Cauchy net in L is w-convergent to an element of L. (By definition w and bw topologies agree in the bounded set E. See Day [16] for a general discussion of the relationship between convergence and closure; in these spaces $x=\lim_{n\in\Delta} F(n)$ if and only if x is in the intersection of the family of sets $wF(A)$, where A runs over the cofinal subsets of Δ.) Note that w- and bw-Cauchy nets in E are the same

too, because they are defined in terms of w- or bw-convergence of nets $(x_n - x_m: (n,m) \in \Delta \times \Delta)$ to 0, and $E-E$ is also bounded.

(3) (a) From (2c) we see that an E in N^* is bw^*-closed if and only if E contains the limit of each bounded w^*-convergent net in E. (b) From Cor. 4,3 it follows that if $X \in N^{**}$ and $d(X, Q(N)) > 0$ and if $H = X^{-1}(c)$, then $bw^*(H) = N^*$. (c) Hence for a *Banach space* B an X in B^{**} is in $Q(B)$ if and only if $X^{-1}(0)$ is bw^*-closed in B^* (Cor. I, 4,2); that is, (d) for hyperplanes in B^*, bw^*-closure is the same as w^*-closure; or for linear functionals on B^*, w^*-continuity is the same as bw^*-continuity.

Lemma 2. *The following definitions all give the bw^*-topology in N^*:*

(i) *A set E is open (closed) if and only if $E \cap U^\pi$ is relatively w^*-open (w^*-closed) in U^π for every ball U about 0 in N.*

(ii) *A set V is a neighborhood of 0 if and only if there is a totally bounded set P in N such that $V \supseteqq P^\pi$.*

(ii_c) *A net is convergent to zero if and only if it is uniformly convergent to zero on each totally bounded subset of N.*

(iii) *A set V is a neighborhood of 0 if and only if there is a compact set A in N such that $V \supseteqq A^\pi$.*

(iii_c) *A net is convergent to 0 if and only if it is uniformly convergent to 0 on each compact subset of N.*

(iv) *V is a neighborhood of 0 if and only if there is a sequence $(y_i, i \in \omega)$ in N such that $\lim_{i \in \omega} y_i = 0$ and $V \supseteqq \bigcap_{i \in \omega} \{y_i\}^\pi$.*

Hence the bw^*-topology in N^* is locally convex.

Proof. A bw^*-closed E obviously satisfies (i) since each U^π is bounded. If E satisfies (i) and C is bounded, take a ball U^π containing C. Then $E \cap C = E \cap U^\pi \cap C$ and $U^\pi \cap w^*(E \cap U^\pi) = E \cap U^\pi$, so $C \cap w^*(E \cap C) \subseteqq C \cap U^\pi \cap w^*(E \cap U^\pi) = C \cap E \cap U^\pi = E \cap C$, and E is bw^*-closed.

To prove that the (i)-topology is as strong as the (ii)-topology it suffices to prove that if P is totally bounded, then each point f where $\sup\{f(x): x \in P\} = 1 - \eta < 1$ is an (i)-interior point of P^π; that is, for each ball U about 0 in N there is a w^*-neighborhood φ^π such that $(f + \varphi^\pi) \cap U^\pi \subseteqq P^\pi \cap U^\pi$.

By total boundedness there is a finite symmetric set ψ such that $\psi + U/2 \supseteqq P$; set $\varphi = 2\psi$ and $E = \varphi^\pi{}_\pi$ to get $P \subseteqq E/2 + U/2 \subseteqq K(E \cup U)$. Hence $P^\pi \supseteqq [K(E \cup U)]^\pi = E^\pi \cap U^\pi = \varphi^\pi \cap U^\pi$. Therefore $P^\pi \cap U^\pi \supseteqq \varphi^\pi \cap U^\pi$. Using this result for $V = \eta U/2$, we see that there is φ such that if $|h(\varphi)| \leqq \eta$ and $h \in 2U^\pi$, then $|h(P)| \leqq \eta$, so $(f + (\varphi/\eta)^\pi) \cap U^\pi \subseteqq f + (\varphi^\pi \cap 2U^\pi) \subseteqq P^\pi$. Hence the set $\bigcup_{0<\eta<1} \eta P^\pi$ is (i)-open and contained in P^π.

(ii) and (ii_c) are equivalent; this is most easily seen from the observation that, because multiples of totally bounded sets are also totally bounded, (ii_c) is equivalent to:

(ii′) A net $(f_n, n\in\Delta)$ is convergent to zero if and only if for each totally bounded set P in L the net (f_n) is ultimately in P^π.

The (ii)-topology is as strong as the (iii)-topology because the compact sets are totally bounded. (iii) is equivalent to (iii$_c$) by the argument used just above.

The (iii)-topology is at least as strong as the (iv)-topology because each (iv)-neighborhood of 0 is a (iii)-neighborhood of 0, for $\{0\}^\pi = N^*$ and $\{0\}\cup\{y_i : i\in\omega\}$ is compact if $\lim_{i\in\omega} y_i = 0$.

To complete the proof let V be a (i)-open set about 0, let U_n be the ball $\{x: \|x\|\leq 1/n\}$ and let φ_1 be a finite set in N such that $\varphi_1^\pi\cap U_2^\pi \subseteqq V$. Then there exists a finite set φ_2 such that $\varphi_2\subseteqq U_2$ and $\varphi_1^\pi\cap\varphi_2^\pi\cap U_3^\pi\subseteqq V$. For should such a φ_2 fail to exist, let Φ be the system of finite subsets of U_2 ordered by inclusion and for φ in Φ let f_φ be any element of $(\varphi_1^\pi\cap\varphi^\pi\cap U_3^\pi)\setminus V$. By w^*-compactness of $U_3^\pi\setminus V$ some subnet of (f_φ) converges to an f which is in all φ^π as well as in $\varphi_1^\pi\cap U_3^\pi\setminus V$. But $\bigcap_{\varphi\in\Phi}\varphi^\pi = U_2^\pi$, and by the choice of φ_1, $(\varphi_1^\pi\cap U_2^\pi)\setminus V=\emptyset$ so there can be no such point as f. This contradiction shows that φ_2 exists; in a similar way we choose by induction φ_n, finite subsets of U_n, such that $(\varphi_1\cup\varphi_2\cup\cdots\cup\varphi_n)^\pi\cap U_{n+1}{}^\pi\subseteqq V$. Enumerating the union of the φ_n yields a sequence $(y_i, i\in\omega)$ such that $\|y_i\|\to 0$ and $\bigcap_{i\in\omega}\{y_i\}^\pi\subseteqq V$. Hence the (iv)-topology is as strong as that of (i).

As (2) shows, w^* and bw^* topologies are different in infinite-dimensional spaces; however the *convex* sets in the conjugate space of a *complete* normed space are better behaved.

Theorem 5 [Kreĭn-Šmul'yan]. *If B is a complete normed space and if K is a convex set in B^*, then $bw^*(K)=w^*(K)$; that is,* (see (3c)) *K is w^*-closed if and only if for all closed balls U in N^*, $K\cap U$ is w^*-closed.*

Proof. Always $w^*(K)\supseteqq bw^*(K)$. For this partial converse, suppose that K is convex and that $g\in bw^*(K)$. Then by Lemma 2 there exists a convex bw^*-neighborhood V of 0 such that $g+V$ is disjoint from $bw^*(K)$. By the separation theorem there is a bw^*-continuous X in $B^{*\#}$ such that $X(g)>\inf\{X(f): f\in V+g\}\geqq\sup\{X(f): f\in K\}$. But (3c) asserts that a bw^*-closed hyperplane is w^*-closed; that is, that X is w^*-continuous. Hence $g\notin w^*(K)$.

[Pryce (letter) provides a direct proof with no use of the bw^*-topology of Th. 5 by finding a set C such that $C^\pi = K$. Let $C_n=[K\cap 2^{n+1}U^\pi]_\pi$; take polars up and down to show that $C_n\subseteqq C_{n+1}+2^{-n}U$. Setting $C=\bigcap_{n\in\omega} C_n$, Pryce proves C non-empty and $C\subseteqq C_{n+1}+2^{-n+1}U$. Since for $\varepsilon>0$, $E+F\subseteqq(1+\varepsilon)co(E\cup\varepsilon^{-1}F)$, $C\subseteqq C_n\subseteqq(1+\varepsilon)(C\cup 2^{n-1}\varepsilon U)$ so for all $\varepsilon>0$, $C^\pi \supseteqq K_n \supseteqq (1+\varepsilon)^{-1}(C^\pi \cap 2^{n-1}\varepsilon U)$. Taking union on n, $C^\pi \supseteqq K \supseteqq (1+\varepsilon)^{-1} C^\pi$ for all $\varepsilon>0$, so $C^\pi = K$.]

(4) (a) In a norm-bounded set E in N^* the w^*-topology determined by any dense linear subset N_0 on N is equivalent to that determined by all of N. (b) If N is infinite-dimensional, a neighborhood basis at 0 for the w^*-topology in any ball about 0 is given by the sets σ^π/(number of elements in σ), where σ runs over the set of all finite subsets of some fundamental set S in N. (c) Hence the w^*-uniformity in each bounded set in N^* has a denumerable base when N is separable (Kelley, p. 164); hence (d) *Each w^*-topologized ball in N^* is a compact metrizable space when N is separable.* [To prove metrizability directly, take a sequence (x_i) dense in U; then $d(f,g)=\sum_i |f(x_i)-g(x_i)|/2^i$ is a metric in every bounded part of N^*.]

(5) (a) If N_0 is a dense linear subset of N and if $\mathfrak{i}$ is the identity function from N_0 into N, then $\mathfrak{i}^*$ is an isometry and a $b\,w^*$-homeomorphism between N^* and N_0^*. (b) Hence, if B is the completion of N, the $b\,w^*$-topology of N^* is the same as that of B^*. (c) The completion of N is isometric to the subspace of N^{**} consisting of all $b\,w^*$-continuous linear functionals on N^*. (d) If $N_0 \neq N$, then $\mathfrak{i}^*$ is continuous but not interior in the w^*-topology.

Lemma 3. *N^* is $b\,w^*$-complete. Hence if (x_n) is a $b\,w$-Cauchy net in N, then $(Q\,x_n)$ is a $b\,w^*$-convergent net in N^{**}.*

Proof. Let $(f_n, n\in\Delta)$ be a $b\,w^*$-Cauchy net in N^*; then (f_n) is also a w^*-Cauchy net, so $f(x)=\lim_n f_n(x)$ exists for each x in N and $f=b\,w^*\text{-}\lim_n f_n$. Also, by I, 4, (3c), if $f\in L^\#$. But if $f\notin L^*$, there exist x_k with $\|x_k\|<1/k$ and $f(x_k)>k$. Then (x_k) converges to zero so $(x_k)^\pi$ is a $b\,w^*$-neighborhood of 0 in N^*. Then (f_n) must converge uniformly on $\{x_k : k\in\omega\}$ so f must be boundet there, a contradiction which proves that $f\in N^*$; that is, N^* is $b\,w^*$-complete.

For the second conclusion, each $b\,w^*$-neighborhood P^π of 0 in N^{**} determines a $b\,w$-neighborhood P_π of 0 in N so $(Q\,x_n)$ is $b\,w^*$-Cauchy if (x_n) is $b\,w$-Cauchy. But N^{**} is also $b\,w^*$-complete, so $(Q\,x_n)$ has a $b\,w^*$-limit in N^{**}.

(6) The $e\,w^*$-topology (Defn. III, 1,2) gives a more general setting for use of equicontinuity. The preceding results extend to a *metrisable* LCS M. (a) *Lemma* 2 *still holds;* the only change needed is in the last section of the proof where (U/n) is replaced by a sequence (U_n) of convex neighborhoods of 0 which form a basis of the 0-neighborhoods in M.

Banach-Dieudonné Theorem. *If M is a metrisable* LCS, *then the $e\,w^*$-topology is a polar set topology $\mathfrak{E}^\pi$, where $\mathfrak{E}$ may be the family of* *totally bounded sets, or of compact sets, or of sequences converging to zero, in M.*

(b) Call a *complete* locally convex metrisable space a *Fréchet space* (Bourbaki's terminology; Banach's *space of type* (F) need not be locally convex.) The Kreĭn-Šmul'yan theorem extends also. *If M is a Frechét space and if K is a convex subset of M^*, then* $e w^*(K) = w^*(K)$. The proof is as before, using (a) in place of Lemma 2 and Thm. III, 1,1 in place of (3c).

(c) Lemma 3 also extends. *If M is a metrizable* LCS, *then M^* is $e w^*$-complete.* The modification of the proof is like that needed for (a).

(7) Rubel and Shields consider the Banach space $H^\infty(G)$ of all analytic functions bounded on the plane region G with supremum norm. They show that $H^\infty(G)$ is like the conjugate space of a separable quotient space of the space of Borel measures on G, so w^* and $b w^*$ topologies can be discussed in $H^\infty(G)$. Rubel and Ryff show that the $b w^*$-topology here agrees with the topology τ which is the strongest topology in which each sequence converges if and only if it is uniformly bounded and convergent at each point of G. They show also that $b w^*$ agrees in $H^\infty(G)$ with the strict topology, which is the locally convex topology defined by the prenorms $p_k(f) = \sup\{|f(z)k(z)| : z \in G\}$, where $k \in C_0(G)$, the set of continuous functions on G which vanish at infinity.

Chapter III. Completeness, Compactness, and Reflexivity

§ 1. Completeness in a Linear Topological Space

In a metric space we have characterized completeness in terms of sequences; in general this does not suffice and we need to use—

Definition 1. A subset E of an LTS L is called *complete* if every Cauchy net $(x_n, n\in\Delta)$ in E has a limit in E; that is, whenever $\lim_{(m,n)\in\Delta\times\Delta}(x_n - x_m)=0$, then x exists in E such that $\lim_{n\in\Delta} x_n = x$. L itself is called *quasi-complete* [*topologically complete*] whenever every closed bounded [every closed totally bounded] set in L is complete.

In Banach spaces completeness and category arguments worked together to yield, for example, the interior mapping theorem. To exploit only the completeness we begin by defining an analogue of the bw^*-topology; the most complete reference for this is the thesis of Wheeler [1].

Definition 2. The *equicontinuous weak** (ew^*-) *topology* in the conjugate of an LCS L is the strongest topology in L^* which agrees with the w^*-topology on every equicontinuous set in L^*; alternatively, a set E is ew^*-open if and only if for every convex symmetric 0-neighborhood U in L, $E\cap U^\pi$ is relatively w^*-open in U^π.

(1) Let L be an LCS. (a) A *convex* set V in L^* is an ew^*-neighborhood of 0 in L^* if and only if there is for each convex symmetric 0-neighborhood U in L a finite subset φ of L such that $\varphi^\pi\cap U^\pi\subseteqq V$. (This is a result of Wheeler [1, Theorem II, 5] which fails if V is not assumed convex, although the earlier editions of this book mistakenly asserted it in general.) (b) By I, 4, (11d) this implies that if V is a convex ew^*-neighborhood of 0 in L^*, then V_π is totally bounded. (c) If W is an ew^*-neighborhood of 0 in L^*, then its convex hull V is also an ew^*-neighborhood of 0 in L^*, so $W_\pi = V_\pi$ is also totally bounded. (d) Hence if W is an ew^*-neighborhood of 0 in L^* and if L is quasi- (or even topologically) complete, then W_π is compact in L. (This proof requires a standard theorem on uniform structures: Every complete totally bounded set in a uniform structure is compact. See Kelley, p. 198.) (e) If E is totally bounded in L,

then $(E \cup -E)^\pi$ is a convex ew^*-neighborhood of 0 in L^*; by (b) $E^\pi{}_\pi \subseteqq (E \cup -E)^\pi{}_\pi$ is totally bounded in L.

(2) (a) In terms of convergence, E is ew^*-closed if and only if each equicontinuous w^*-convergent net in E has its limit in E. [This follows from the facts that each U^π is an equicontinuous set of functions on L and that each equicontinuous set from L^* is contained in some positive multiple of some U^π.] (b) If L is a normed space, then the ew^*-topology coincides with the bw^*-topology of Def. II, 5,2.

(3) (a) Convergence of a net $(x_n, n \in \Delta)$ in L (with the topology given by $\mathfrak{U}$, of course) is equivalent to convergence of the net of functions $(Q x_n, n \in \Delta)$ uniformly on each set U^π in L^*. (b) Hence Cauchy nets in L are those nets whose images under Q satisfy a uniform Cauchy condition on each U^π. (c) Each X in $L^{*\#}$ which is the limit of such a Cauchy net $(Q x_n, n \in \Delta)$ is w^*-continuous on each U^π.

Ptak [1] proved the converse of (3c); see also Grothendieck [1] and Roberts.

Theorem 1. *The completion of an* LCS *L is isomorphic to the set $L^\wedge$ of all those functions in $L^{*\#}$ which are weak*-continuous on each U^π, U a neighborhood of* 0 *in L; i.e., to the set of ew^*-continuous linear functionals on L^*.*

Proof. Take X in $L^{*\#}$ continuous on each U^π. The main problem is to approximate X uniformly closely on each U^π by an element $Q x$ with x in L.

Take p to be the Minkowski functional of U; then p is a continuous pre-norm in L, and the set $L_0 = \{x : p(x) = 0\}$ is a closed linear subspace of L. Then $N = L/L_0$ is normed by p (but this norm topology in N may be much weaker than the factor space topology in L/L_0). If T is the natural mapping of L onto L/L_0, then T^* maps N^* into $L_0^\perp \cap L^*$, T^* is a homeomorphism in the w^*-topologies, and T^* carries the unit ball S of N^* onto U^π. Hence X determines a bw^*-continuous linear functional $\xi = X \circ T^*$ on N^*. If N were complete, II, 5, (3d) would assert that ξ is a $Q' y$, y in N; lacking that assumption all that can be asserted is that ξ can be approximated on S uniformly within $1/n$ by a $Q' y$. Then there is in L an element x_{Un} such that $y = T x_{Un}$, so $Q x_{Un} - X$ is uniformly $\leqq 1/n$ on U^π. The net $(Q x_{Un}, (U, n) \in \mathfrak{U} \times \omega)$ converges to X in the topology of uniform convergence on sets U^π. Hence (x_{Un}) is a Cauchy net in L; this with (3) proves the theorem.

Corollary 1. *If L is an* LCS *with neighborhood basis $\mathfrak{U}$ of* 0, *then:* (a) *An X in $L^{*\#}$ is in the completion of L if and only if Z, the set of zeros of X, is ew^*-closed.* (b) *An X in $L^{*\#}$ is in the completion of L if and only if X is ew^*-continuous on L^*.* (c) *L is complete if and only if in L^* ew^*-closure of hyperplanes in equivalent to w^*-closure of hyper-*

planes. (d) *L is complete if and only if ew^*-continuity and w^*-continuity are equivalent for linear functionals on L^*.*

The chief difficulty with the ew^*-topology is that it need not be locally convex or even a topology in which addition is continuous in two variables.

(4) (a) In $(R^S)^*$ the w^*-topology is the weak topology related to the convex core topology. (b) By I, 4, (13b) and Cor. I, 6, 6 every linear subspace of $(R^S)^*$ is w^*-closed. (A space with this property of R^S is called by Collins "fully complete".)

(5) L' be an LCS and let L be the space L' retopologized with the weak topology, then: (a) Each $U^\pi = (\varphi_\pi)^\pi$ is a finite-dimensional compact set in L^*. (b) U^π is a neighborhood of 0 in the finite-dimensional space $\bigcup_{n\in\omega} n\,U^\pi$. (c) Every linear functional on L^* is ew^*-continuous. (d) The completion of L is all of $L^{*\#}$. (e) L is complete if and only if there is an index set S such that L is isomorphic to the topological product space R^S. (f) No infinite-dimensional normed space can be complete in the weak topology [Taylor].

(6) If M is a metrizable linear space, then L^* is ew^*-complete; that is, is w^*-quasi-complete. (See II, 5, (6c).) Wheeler [1] gives many old and new examples of spaces for which L^* is not ew^*-complete; for example, if L is a linear space of uncountable dimension and if L is given the finest locally convex topology, then the ew^*-topology for L^* is not completely regular, so there is no uniformity compatible with the topology so no way to complete L.)

(7) Ptak [1] has analysed Banach's proof of the interior mapping theorem in the following terms. A linear continuous T from one LCS L into another L' is called *almost-interior* if T carries each open set in L to a somewhere dense set in L'. Then Banach's proof of the interior mapping theorem, which has been written in II, § 3 as a proof of the closed graph theorem, divides into two steps; the first step uses a category argument in L' which proves that T is almost-interior. The second step uses completeness in L to show that an almost-interior T is interior. Ptak shows how the second argument is related to completeness by comparing the first of the conditions of the next theorem with the condition, equivalent to completeness of L, that ew^*-closed hyperplanes are w^*-closed.

Theorem 2. *The following two conditions on an* LCS *L are equivalent: (B_1) Every ew^*-closed linear subset of L^* is w^*-closed. (B_2) Every almost-interior linear continuous T from L into an* LCS *L' is interior.*

Proof. To show that (B_1) implies (B_2), let T be almost-interior. For each U in $\mathfrak{U}$ there is a U' in $\mathfrak{U}'$ such that $w(TU)) \supseteqq U'$; then $T(U)^\pi \subseteqq U'^\pi$ and $U^\pi \cap T^*(L'^*) = U^\pi \cap T^*(U'^\pi)$ is the intersection of w^*-compact sets and hence is w^*-closed. (B_1) implies that $T^*(L'^*)$ is w^*-closed; hence, by I, 5, (7), it is all of $T^{-1}(0)^\perp \cap L^*$. This implies that $T(U)^\pi_\pi = T(U^\pi_\pi)$. If (Tx_n) is a Cauchy net in L', this result (for all U) implies that (x_n) is a Cauchy net in L. By Cor. 1, (c), (B_1) implies that L is complete, so each element of U' is in $T(U)$; that is, T is interior.

If on the other hand (B_2) holds and E is an ew^*-closed subspace of L^*, let $F = E_\perp$, let T be the natural mapping from L onto the factor space L/F, and create L' by topologizing L/F with the set of neighborhoods $U' = (T^{*-1}(E \cap U^\pi))_\pi$. If in $(L/F)^*$ we set $E' = T^{*-1}(E)$ and use bw^*-closure of E, then we have $U' = (E' \cap (T(U)^\pi)_\pi$. Then $L'^* = E'$, and for each U the L'-closure of $T(U)$ is the weak-E'-closure of $T(U)$ $= (E' \cap T(U)^\pi)_\pi$, an L'-neighborhood of 0, so T is almost interior. Assumption (B_2) then makes T interior, so the topology of L is that of L/F, so $E' = L'^* = (L/F)^*$; that is, $E = T^*(E') = T^*((L/F)^*) = F^\perp$, which is a w^*-closed subspace of L^*.

In the same paper Ptak also gives examples of continuous-function spaces with the compact-open topology (i) one of which is complete and does not satisfy (B_2), and (ii) another of which is topologically complete but not complete.

(8) (a) For a normed space N the ew^*-topology is the bw^*-topology of II, § 5. (b) Hence Theorem 2 with the Kreĭn-Šmul'yan Theorem II, 5,5 or II, 5, (6b) gives a proof that *every almost-interior continuous linear operator carrying a* Fréchet *space into an* LCS *is interior.*

(9) Wheeler [1, 2] defined two topologies in an LCS L. If $\mathfrak{B}$ is the family of bounded sets in L, then the ew^*-topology of $(L^*, \mathfrak{B}^\pi)^*$ determines a relative topology in $Q(L)$; this topology carried back to L by Q^{-1} is *the ew-topology in L.* The finest topology which agrees with the w-topology of L on all sets in $\mathfrak{B}$ defines *the bw-topology in L.* (a) If $B \in \mathfrak{B}$, then $B \subseteqq B^\pi_\pi \in \mathfrak{B}$ and $Q(B) \subseteqq Q(B^\pi_\pi) \subseteqq B^{\pi\pi}$, so bw has at least as many closed sets as ew in L. (b) Wheeler [1], p. 97, gave an example, quoted in II, 5, (2b), of a set in c_0 for which $ew(A) \neq bw(A)$. The same space shows that bw need be neither locally convex nor a vector topology for L. (Wheeler [1], p. 98.) (c) Wheeler [1], Chapter V, also considers the strongest locally convex topologies, cew^*, cew, cbw, which agree with w^* or w on appropriate subsets. Part of his motive is to extend to some larger class of spaces the characterization in Thm. 4, 2, (Q) of reflexivity of a Banach space.

(10) Let L be an infinite-dimensional vector space, let $L^* = L^\#$, and use in L the w-topology. (a) By (5), M, the completion of L, is isomorphic to $L^{*\#}$ in its w^*-topology; hence M is larger than L. (b) E bounded in L implies E is finite-dimensional. (c) Hence the closure of E in L is compact. (d) Hence every bounded closed subset of L is closed in M. (e) L is ew-closed in M but is w-dense.

(11) Collins gave a simple example of a space whose ew^*-topology is not locally convex. Let B be any infinite-dimensional Banach space and let L have the vector structure of B and the convex-core topology. Then: (a) $L^* = B^\# + B^*$. (b) B^* is w^*-dense in L^*. (c) If U is a neighborhood of 0 in L, then $U^\pi \cap B^*$ is w^*-bounded; therefore it is norm-bounded; therefore it is w^*-compact. Hence B^* is ew^*-closed in L^*. (d) If the ew^*-topology were locally convex in L^*, then each g not in B^* could be separated from B^* by an ew^*-closed hyperplane in L^*, which, by completeness of L, would be w^*-closed; this would imply that B^* is w^*-closed, contradicting (a) and (b).

(12) Porta discusses in each LCS L the finest locally convex topology $l(L)$ in which E is closed in $l(L)$ if and only if for each compact set K in L, $E \cap K$ is closed in the original topology. If L is (B^*, w^*), then $l(L)$ is the bw^* topology in B^*. The topology $l(L)$ is determined by the family of all pre-norms in L which are continuous on compact subsets of L, and $l(L)$ is the finest locally convex topology with the same compact subsets as L. Porta gives conditions, such as metrizability, sufficient for $l(E) = E$, discusses permanence properties, and relations with other structures, such as sequential topologies, derived from L.

§ 2. Compactness

In a general topological space several variations are available for the notion of compactness; let us list some of these until we have proved that in norm and in weak topologies of normed spaces all these agree.

Let L be a topological space (Hausdorff be it understood) and let E be a subset of L. E is called *compact* [symbol: (C)] if every net $(x_n, n \in \Delta)$ in E has a subnet converging to a point x of E; E is *countably compact* [(cC)] if every countable net (or every sequence) in E has a *subnet* converging to a point x of E; E is *sequentially compact* [$(\omega$C)] if every sequence in E has a *subsequence* converging to a point x of E. The corresponding definitions of *relative compactness* require that x be in L, not necessarily in E.

(1) (a) (C) or (ωC) implies (cC). (b) The corresponding relations are also true for the corresponding relative properties. (c) No other such implication holds in general. The countable ordinals with the order topology are sequentially compact but not compact, while a product of continuum-many intervals is compact (Tychonov's theorem) but is not sequentially compact. (d) Embedding a completely regular space T homeomorphically into $C(T)^*$, as in Chapter V, § 3, it follows that for subsets of N^* in the w^*-topology no better general relationships can hold.

(2) In a metric space all three properties are equivalent.

(3) In an LTS countable compactness implies total boundedness.

The conditions used in the next theorem are to be found in Šmul'yan [1,2,8] and Gantmaher and Šmul'yan.

Theorem 1. *In a normed linear space the following properties of a set E are equivalent to each other; the corresponding relative properties are also equivalent to each other:*

(w C) *E is weakly compact.*

(w ω C) *E is weakly sequentially compact.*

(w c C) *E is weakly countably compact.*

(Γ) *For each sequence* $(x_n, n\in\omega)$ *in E there is x in E such that for each* f *in* N^*, $\liminf_{n\in\omega} f(x_n) \leq f(x) \leq \limsup_{n\in\omega} f(x_n)$.

(Š) *If* $(K_n, n\in\omega)$ *is a sequence of closed convex sets in N such that for each n,* $K_n \supseteqq K_{n+1}$ *and* $K_n \cap E \neq 0$, *then* $\bigcap_{n\in\omega} K_n \cap E \neq \emptyset$.

Proof. (w C) or (w ω C) implies (w c C). (w c C) implies (Γ) because the limit x of any w-convergent subnet of (x_n) will serve. If K_n are given and x_n is any point of K_n, then any x satisfying (Γ) will, by Mazur's theorem in the form of Cor. I, 6, 2, be in all K_n, so (Γ) implies (S).

The hard part of the proof is to show that (Š) implies (w C) and (w ω C); the present proof grew from (i) Kaplansky's verbal proof in 1948-9 that in Hilbert space (w ω C) implies that E is w-closed. (ii) Thm. III, 2,4 of the first version of this book which extended that to all normed spaces, and (iii) the proof of Whitley [1] of Eberlein's theorem that (w ω C) implies (w C). The crux is: *If E satisfies* (Š), *then* $Q(E)$ *is* w^*-*closed*. Since (Š) trivially implies that E is bounded, $w^*(Q(E))$ will be w^*-compact which will make E w-compact.

Take X in $w^*(Q(E))$. There is f_1 of norm one in N^* such that $X(f_1) \geq \|X\|/2$. Then there is x_1 in E for which $|X(f_1) - f_1(x_1)| < 1/2$. In this way we use II, 4, (3a) raised one level alternately with the fact that X is in $w^*(Q(E))$ to get a sequence of finite subsets $\varphi_1 \subseteqq \varphi_2 \subseteqq \cdots \subseteqq \varphi_n \cdots$ of the unit sphere in N^* and a sequence (x_j) of elements of E such that for all Y in the linear hull of $\{X, Q x_1, \ldots, Q x_{n-1}\}$ we have (i) $\|Y\| < ((n+1)/n) \max\{Y(f): f \in \varphi_n\}$ and (ii) $|X(f) - f(x_n)| < 1/n$ if $f \in \varphi_n$. Then for every f in $\bigcup_{n\in\omega} \varphi_n$, $\lim_j f(x_j) = X(f)$ and for every Y in L, the closed linear hull of $\{X, Q x_1, \ldots, Q x_n, \ldots\}$ we have $\|Y\| \leq \sup\left\{Y(f): f \in \bigcup_n \varphi_n\right\}$. Let K_j be the closed convex hull of $\{x_i: i \geq j\}$. Then by (Š) there is a point x in $E \cap \bigcap_{j\in\omega} K_j$ and $Q x \in L$. But x in $\bigcap_{j\in\omega} K_j$ implies that $X(f) = f(x)$ for every f in $\bigcup_n \varphi_n$. But this set is total over L so $X = Q x$. This shows that $w^*(Q(E)) = Q(E)$; hence $Q(E)$ is w^*-compact and E is w-compact if E is (Š).

Also in the preceding paragraph, $x = w\text{-}\lim_j x_j$, for if not there would be f_0 and $\varepsilon > 0$ such that $f_0(x_{j_k}) \geq f_0(x) + \varepsilon$ for all k in ω, so $E \cap E_n \cap \{y: f_0(y) \geq f_0(x) + \varepsilon\}$ would be non-empty for all n, so, by (Š), there would be a z in the intersection of all these sets, so $Qz \in L$ and then $Qz = X = Qx$, contrary to $f_0(z) \geq f_0(x) + \varepsilon$. Now if (a_k) is any sequence in E and X any w^*-cluster point of $(Q a_k)$, we could have chosen the x_i to be a subsequence of (a_i). Hence E is (wωC) if it is (Š).

This last proof (that (wC) implies (wωC)) does not work out in a general LCS because this proof uses: (1) Each closed separable subspace of L^{**} has a countable total set in L^* (a property of normed spaces by II, 1, (4e)) and (2) the version of compactness used, (Š) is this case, is inherited from E to each $E \cap A$ where A is any closed linear subspace of L. In an LCS with (1) we can push through the proof that (Š) implies (wωC); hence if we hope to prove in a space where (wC) does not imply (wωC) that (wc$C$) implies (wC) (for example $L = m(\omega)^*$ with the w^*-topology, E = unit ball) we cannot use this method. We describe next a method due to James that uses (2) without (1) (Theorem 2) and one basically due to Eberlein [1] modified by Ptak and Day which does not need (1) or (2) so proves, for example, that weak pseudocompactness implies (wC).

Let us now consider properties inherited by all subsets; since these cannot imply closedness, they can at best characterize some sort of relative compactness. Some of these conditions were used by Grothendieck [2], James [4,5], and by Eberlein [1] who first attacked $w^*(Q(E))$ as a means to prove weak compactness.

Lemma 1. *For a bounded set E in an* LCS *L each condition in the following list implies the next:*

(G) *If* $(f_i, i \in \omega)$ *is an equicontinuous sequence of elements of* L^* *and if* $(x_j, j \in \omega)$ *is a sequence from E such that all single and both iterated limits exist, then* $\lim_i \lim_j f_i(x_j) = \lim_j \lim_i (x_j)$.

(Ja) *There do not exist a number* $\theta > 0$, *an equicontinuous sequence* (f_i) *in* L^*, *and a sequence* (x_j) *in E such that* $f_i(x_j) > \theta$ *when* $i \leq j$ *and* $f_i(x_j) = 0$ *when* $i > j$.

(E_r) $w^*(Q(E)) \subseteqq L^\wedge$, *the canonical completion of L in* $L^{*\#}$. (See Thm. 1,1.)

Proof. (G) implies (Ja). If (Ja) fails and θ, (f_i), and (x_j) do exist, by equicontinuity of the (f_i) and boundedness of the (x_j) the set of numbers $\{f_i(x_j): i, j \in \omega\}$ is bounded, hence conditionally compact in R. By the Cantor diagonal process, subsequences exist with the hypotheses of (G), but (G) must fail because $\lim_i \lim_j f_i(x_j) \geq \theta > 0 = \lim_j \lim_i f_i(x_j)$.

(J a) implies (E_r). If (E_r) fails, take X in $w^*(Q(E))\setminus L^\wedge$. Embed L in a complete LCS L' and let W be the usual isomorphism of II, 3, (11f) of L' into a topological product of Banach spaces, $\prod_{s\in S} B_s = M$, and let W_s be the sth coordinate of W. Then by II, 3, (11) $M^* = \sum_{s\in S} B_s^*$ and $Y = W^{**}(X)$ is in $\prod_{s\in S} B_s^{**}$. Since $X = w^*\text{-}\lim_n Q x_n$ for some net $(x_n, n\in\Delta)$ in E, for each s in S the net $(W_s x_n)$ is a bounded w-Cauchy net in B_s and $Q_s W_s x_n$ has some w^*-limit Y_s in B_s^{**}; this element is precisely the sth coordinate of $Y = W^{**} X$.

In the case in which for each s there is a y_s in B_s with $Y_s = Q y_s$, then $w\text{-}\lim_n W_s x_n = y_s$ for all s and, if y is the element of M with coordinates y_s, then $y = w\text{-}\lim_n W x_n$. Hence y is in the strong closure of $W(L)$ which is $W(L')$. Then there is an x in L' such that $y = Wx$ and then $X = Q x$, so X is in $L^\wedge$ by Theorem 1.1. But we assumed $X \in L^\wedge$, so this case does not occur and there is at least one s for which $Y_s \notin Q_s B_s$; that is $d(Y_s, Q_s(B_s)) = d > 0$. That is, to prove that (J a) implies (E_r) in every LCS we need only prove it for Banach spaces, and we drop the subscript s temporarily.

We shall construct a sequence contradicting (J a) for each θ with $0 < \theta < d$. Let V be the unit ball of B^*. Then there is g_1 in V such that $Y(g_1) > \theta$, because $\|Y\| \geq d > \theta$. Since $Y \in w^*(Q(E))$, there is u_1 in E with $g_1(u_1)$ arbitrarily near $Y(g_1)$ so take $g_1(u_1) = \theta$. By Cor. II, 4,2 there is g_2 in V for which $Y(g_2) > \theta$ and $g_2(x_1) = 0$. Because $Y \in w^*(Q(E))$ again there is u_2 in E with $g_i(u_2) = \theta$ for each $i \leq 2$. Alternately using Y in $w^*(Q(E))$ and Cor. II, 4,2, we construct sequences (g_i) in V and (u_j) in E so that $g_k(u_j) = \theta$ when $i \leqq j$ and $g_i(u_j) = 0$ when $i > j$.

This shows that if (E_r) fails in a Banach space then (J a) fails too. To return to L, let $f_i = W_s^*(g_i)$ and choose x_j in E so that $W_s(x_j) = u_j$. These sequences contradict (J a) for L.

Theorem 2. *If E is a bounded subset of an* LCS *L and if for any reason $w^*(Q(E)) \subseteqq Q(E)$, then $w(E) = Q^{-1}(w^*(Q(E)))$ is w-compact. Hence if L is complete or even quasi-complete, then the following conditions on a bounded set E are equivalent:*

($\mathrm{w\,C}_r$) *$w(E)$ is* (w C), *that is, E is relatively w-compact, or every net in E has a w-convergent subnet.*

($\mathrm{w\,c\,C}_r$) *$w(E)$ is* (w c C), *that is, every sequence in E has a w-convergent subnet.*

(Γ_r) *For each sequence (x_j) in E there is x in $w(E)$ such that $\liminf_j f(x_j) \leq f(x) \leq \limsup_j f(x_j)$ for all f in L^*.*

($\check{\mathrm{S}}_r$) *For each decreasing sequence (K_j) of closed convex sets in L such that each K_j meets E there is an x in the intersection of $w(E)$ and all K_j.*

(G), (J a), *and* (E_r) *as in Lemma* 1.

(E) $w^*(Q(E)) \subseteq Q(L)$.

(Note that the first four conditions are already strong enough to force boundedness on E, but any line L is an example of a set satisfying the last four conditions but not bounded nor, of course, compact.)

Proof. The implications (w C_r) implies (w c C_r) implies (Γ_r) implies ($\check{S}_r$) have proofs like those of Theorem 1; (G) implies (J a) implies (E_r) by Lemma 1.

To see that ($\check{S}_r$) implies (G) take (f_i) and (x_j) satisfying the hypotheses of (G) and let K_n be the closed convex hull of $\{x_j : j \leq n\}$. Then there is an x in $\bigcap_n K_n$. Then for each n in ω there is a convex combination y_n of the $x_j, j \geq n$, with $\|y_n - x\| < 1/n$. Since the y_n are averages of elements far out in the sequence (x_j), any convergence is inherited; that is, for each i, $\lim_n f_i(y_n) = \lim_j f_i(x_j)$. But $\lim_n |f_i(y_n) - f_i(x)| = 0$ and the convergence is uniform on i, so i, n limits can be interchanged to get $\lim_i \lim_j f_i(x_j) = \lim_i \lim_n f_i(y_n) = \lim_n \lim_i f_i(y_n)$. But $\lim_j \lim_i f_i(x_j)$ exists and each $\lim_i f_i(y_n)$ is an average of some $\lim_i f_i(x_j)$ so these limits also equal $\lim_j \lim_i f_i(x_j)$ and (G) holds.

w-quasi-completeness implies that (E_r) implies (E). (E) implies (w C_r).

Definition 1. A set E in a topological space L is [*relatively*] *pseudo-compact* (p C) if for every real-valued continuous function f defined but unbounded on E there is a point x in [the closure of] E such that f is unbounded in every neighborhood of x.

(4) In a topological space L a set E is [relatively] (p C) if and only if (a) every continuous real-valued function defined on [the closure of] E is bounded, and if and only if (b) every such function carries [the closure of] E onto a compact subset of R, and if and only if (c) every such function attains its maximum on [the closure of] E.

(5) (a) (c C) implies (p C). (b) Each continuous function carries [relatively] (p C) sets to [relatively] (p C) sets. (c) A [relatively] (p C) set of real numbers is [relatively] compact. (d) A relatively (w p C) set in an LCS L is weakly bounded; hence, by II, 3, (12), is bounded in L.

(6) (a) Following Ptak we consider a weaker condition on $E \subseteq L$.

(Pt) [(Pt_r)]. For each X in $L^{*\#}$ which is in $w^*(Q(E))$ and each sequence (f_i) in L^* there is an x in $E[w(E)]$ such that $X(f_i) = f_i(x)$ for all i in ω.

A still weaker condition is

(Pc_0) [(Pc_{0r})]. For each X in $L^{*\#}$ which is in $w^*(Q(E))$ and each equicontinuous sequence (f_i) in L^* there is an X in E [in $w(E)$] such that $\lim_{i \in \omega} (X(f_i) - f_i(x)) = 0$. (b) (Pt) implies ($Pc_0$) and ($Pt_r$) implies ($Pc_{0r}$).

(c) ($\mathrm{Pc_0}$) implies that much of (Pt) in which the f_i are assumed equicontinuous. (Arrange the f_i with repetitions into a sequence (g_j) where each f_i appears infinitely often.) (d) Hence ($\mathrm{Pc_0}$) implies (Pt) in a normed space.

Lemma 2. *For a bounded set E in an* LCS *L we have:*

$$\begin{array}{ccccccc} (\mathrm{wpC}) & \Rightarrow & (\mathrm{Pt}) & \Rightarrow & (\mathrm{Pc_0}) & & \\ \Downarrow & & \Downarrow & & \Downarrow & & \\ (\mathrm{wpC}_r) & \Rightarrow & (\mathrm{Pt}_r) & \Rightarrow & (\mathrm{Pc}_{0r}) & \Rightarrow & (\mathrm{E}_r) \end{array}$$

Proof. To see that (wpC) implies (Pt) let φ_i be defined on L by $\varphi_i(x)=|X(f_i)-f_i(x)|$; use (5e) to show that $M_i=1+\sup\{\varphi_i(x): x\in E\}$ is finite for each i. Then $\Phi(x)=\sum_{i\in\omega}\varphi_i(x)/2^i M_i$ is the sum of a uniformly convergent series of w-continuous functions on $w(E)$ and is, therefore, w-continuous on $w(E)$. Because $X\in w^*(Q(E))$, for each integer n there is an x_n in E with $|X(f_i)-f_i(x_n)|<1/n$ for all $i<n$. Hence $\inf\{\Phi(x): x\in E\}=0$.

Hence, by (4c), $-\Phi$ attains its maximum 0 at some point x in E; that is, $0=X(f_i)-f_i(x)$ for all i in ω.

(Pt) implies ($\mathrm{Pc_0}$) is purely formal, and the corresponding relative conditions work in the same way. All that is left is to use Eberlein's device to prove that ($\mathrm{Pc_0}$) implies (E_r). Take X in $w^*(Q(E))$; then we want X to be ew^*-continuous, that is, for each convex 0-neighborhood U in L we want to show that $U^\pi\cap X_\perp$ is w^*-closed. U^π is w^*-compact so any g in $w^*(U^\pi\cap\{X\}_\perp)$ is in U^π. Because $X\in w^*(Q(E))$, there exists x_1 in E such that $|X(g)-g(x_1)|<1$; next there is f_1 in $U^\pi\cap\{X\}_\perp$ such that $|f_1(x_1)-g(x_1)|<1/2$; then there is x_2 in E such that $|X(f_1)-f_1(x_2)|<1/3$ and $|X(g)-g(x_2)|<1/4$. Alternately using the conditions $X\in w^*(Q(E))$ and $g\in w^*(U^\pi\cap\{X\}_\perp)$, two sequences can be constructed, (f_i) in $U^\pi\cap\{X\}_\perp$ and (x_j) in E, for which $\lim_{j\in\omega} g(x_j)=X(g)$, $\lim_{i\in\omega} f_i(x_j)=g(x_j)$ and $\lim_{j\in\omega} f_i(x_j)=X(f_i)=0$.

By w^*-compactness of U^π there exists a subnet $(f_{i_m}, m\in\Delta)$ of $(f_i, i\in\omega)$ such that $w^*\text{-}\lim_{m\in\Delta} f_{i_m}$ exists and is an element f of U^π. Let the sequence $(\varphi_i, i\in\omega)$ be defined by $\varphi_{2i-1}=f$, $\varphi_{2i}=f_i$ for each i in ω, and let $(x_{j_n}, n\in\Delta')$ be any subnet of $(x_j, j\in\omega)$ for which $(Q x_{j_n}, n\in\Delta')$ is w^*-convergent to some limit Z. (Such a subnet must exist because $w^*(Q(E))$ is w^*-closed and bounded in $L^{*\#}$.) By ($\mathrm{Pc_0}$) there is a z in E such that $\lim_{i\in\omega}(Z(\varphi_i)-\varphi_i(z))=0$. Then the same limit holds for every subnet of (φ_i); in particular, $Z(f)-f(z)=0$ and

$$\begin{aligned} X(g)&=\lim_{j\in\omega} g(x_j)=\lim_{n\in\Delta'} g(x_{j_n})=\lim_{n\in\Delta'}\Big(\lim_{i\in\omega} f_i(x_{j_n})\Big)\\ &=\lim_{n\in\Delta'}\Big(\lim_{m\in\Delta} f_{i_m}(x_{j_n})\Big)=\lim_{n\in\Delta'} f(x_{j_n})=Z(f)\end{aligned}$$

$$=f(z)=\lim_{m\in\Delta} f_{i_m}(z)=\lim_{m\in\Delta} Z(f_{i_m})$$
$$=\lim_{m\in\Delta}\left(\lim_{n\in\Delta'} f_{i_m}(x_{j_n})\right)=\lim_{m\in\Delta}\left(\lim_{j\in\omega} f_{i_m}(x_j)\right)$$
$$=\lim_{m\in\Delta} X(f_{i_m})=\lim_{m\in\Delta} 0=0.$$

Hence $U^\pi\cap\{X\}_\perp$ is w^*-closed in L^*.

Corollary 1. *If* L *is a quasi-complete* LCS, *then an element* X *of* $L^{*\#}$ *is in* $Q(L)$ *if and only if there is a bounded subset* E *of* L *which satisfies* ($\mathrm{P\,c_0}$) *and is such that* $X\in w^*(Q(E))$.

If $X=Qx$, $E=\{x\}$ will do. If such an E exists, Lemma 2 asserts that X is $e\,w^*$-continuous. Corollary 1.1 (d) asserts that $X\in Q(L)$.

(7) (w p C) trivially implies a condition used successfully in a paper of James [5].

(J) Every f in L^* attains its supremum on E.

The converse is included in James [5], Th. 6: *If a subset* E *of a quasi-complete* LCS L *has property* (J), *then* E *is* ($\mathrm{w\,C_r}$); *that is,* $w(E)$ *is* (w C).

This converse was proved in a sequence of papers of James [3, 4, 5, 10]; the representation of II, 3, (11) reduces this quickly to the Banach space case; we give here the proof for the separable case only.

James's new proof [10] that for separable Banach spaces (J) implies (w C) is wonderfully shortened. Let E be a bounded w-closed set, let $\beta=\sup\{\|x\|:x\in E\}$, and let p be the functional on B^* defined by $p(f)=\sup f(E)$. Then if E is not (wC), as in the proof of $(\mathrm{Pc_0})\Rightarrow(\mathrm{E_r})$, we have:

(a) There is a number θ with $0<\theta<\beta$ and a sequence (f_i) in U^π such that $p(f)\geq\theta$ for all f in the closed convex hull of the (f_i), and (f_i) is w^*-convergent to 0.

[Choose X any element of $w^*(Q(E))\setminus Q(B)$ and $\theta<\|X+Q(B)\|$ (which is $\leq\beta$).] The calculations come in proving that if E satisfies (a), then:

(b) If $\sum_{i\in\omega}\lambda_i=1$ with all $\lambda_i>0$, then there exists a sequence (g_i) and a number α, $\theta\leq\alpha\leq\beta$, such that $p\left(\sum_{i\in\omega}\lambda_i g_i\right)=\alpha$ but for each n in ω, $p\left(\sum_{i\leq n}\lambda_i g_i\right)<\alpha\left(1-\theta\sum_{i\geq n}\lambda_i/\beta\right)$, and (g_i) is w^*-convergent to 0.

Let V_n be the convex hull of the f_i, $i\geq n$ and let $s_n=\sum_{i\geq n}\lambda_i$. Choose positive numbers ε_k so small that $\sum_{i\in\omega}\lambda_i\varepsilon_i/s_{i+i}s_i<1-\theta/\beta$. Define α_n and g_n inductively from $g_0=0$, and, if $g_0,\dots,g_{n-1}$ are known, let

$$\alpha_n=\inf\left\{p\left(\sum_{i\leq n-1}\lambda_i g_i+s_n g\right):g\in V_n\right\},$$

and if α_n is known choose g_n so that

$$p\left(\sum_{i\leq n-1}\lambda_i g_i+s_n g_n\right)<\alpha_n(1+\varepsilon_n).$$

Then, by (a), $\theta \leq \alpha_n \leq \beta$ for each n. Also $\alpha_{n+1} > \alpha_n$, so the α_n increase toward a limit $\alpha \leq \beta$ and $p\left(\sum_{i \in \omega} \lambda_i g_i\right) = \alpha$. To prove the desired inequality,

$$p\Big(\sum_{i \leq n} \lambda_i g_i\Big) \leq \frac{\lambda_n}{s_n} p\Big(\sum_{i \leq n-1} \lambda_i g_i + s_n g_n\Big) + \frac{s_{n+1}}{s_n} p\Big(\sum_{i \leq n-1} \lambda_i g_i\Big)$$

$$\leq s_{n+1} \left[\frac{\lambda_n \alpha_n (1+\varepsilon_n)}{s_{n+1} s_n} + \frac{p\Big(\sum_{i \leq n-1} \lambda_i q_i\Big)}{s_n} \right].$$

Apply this repeatedly with decreasing n to get

$$p\Big(\sum_{i \leq n} \lambda_i g_i\Big) < s_{n+1} \left[\frac{\sum_{i \leq n} \lambda_i \alpha_i (1+\varepsilon_i)}{s_{i+1} s_i} \right].$$

Increase α_k to α, use the defining condition of ε_k, and simplify to get

$$p\Big(\sum_{i \leq n} \lambda_i q_i\Big) < \alpha s_{n+1} \left[\frac{\sum_{i \leq n} \lambda_i}{s_{i+1} s_i} + \frac{\varepsilon_i \lambda_i}{s_{i+1} s_i} \right]$$

$$< \alpha s_{n+1} \left[\sum_{i \leq n} \left(\frac{1}{s_{n+1}} - \frac{1}{s_n} \right) + 1 - \frac{\theta}{\beta} \right].$$

It remains to show that if E has (b) then (J) fails because $\sum_{i \in \omega} \lambda_i g_i$ does not attain its maximum α. Given x in E choose n so large that $g_i(x) < \alpha\theta/\beta$ if $i > n$. Then

$$\sum_{i \in \omega} \lambda_i g_i(x) < \sum_{i \leq n} \lambda_i g_i(x) + \frac{\alpha \theta s_{n+1}}{\beta} < \alpha \left(1 - \frac{\theta s_{n-1}}{\beta}\right) + \frac{\alpha \theta s_{n+1}}{\beta} = \alpha.$$

In non-separable spaces (g_i) cannot always be chosen to be w^*-convergent to 0, so a more complicated (but still vastly improved) calculation and function are needed; see James [10].

This is the end of our long spiral tour through weak compactness.

Theorem 3. *For a bounded, w-closed subset E of a quasi-complete* LCS L (wωC) *implies* (wC), *and all the conditions of this section except* (wωC) *are equivalent. For a bounded set E in a quasi-complete space the relative conditions are equivalent to each other and to: w(E) is* (wC). *In a Banach space, or even in a Fréchet space, these conditions are also equivalent to* (wωC), *or* (wωC$_r$), *respectively.*

(8) Ptak [4] gives an example of a set with (Pt) which is not w-closed; it can be seen not to be pseudocompact. Let L be $c_0(S)$, with S uncountable. The w-topology in any bounded set is that of coordinatewise convergence. Let M be the set of basis vectors $\{\delta_s, s \in S\}$, in $c_0(S)$, let B=closed convex hull of M. (a) $0 \in w(M)$ and $M \cup \{0\}$ is w-compact.

(b) B is w-compact and $B=\left\{x: x(s)\geqq 0 \text{ for all } s \text{ in } S \text{ and } \sum_{s\in S} x(s)\leqq 1\right\}$.

(c) Set $E=\left\{x: x\in B \text{ and } \sum_{s\in S} x(s)=1\right\}$. Then E has (Pt), is dense in B, and is not w-pseudocompact nor w-closed.

Pertinent papers include Eberlein [1], Grothendieck [2] (perhaps the most general study near this field), Ptak [2,3,4,5] and Dieudonné [3]. Dieudonné proves directly that (Š) implies (wC) for a convex closed subset of a quasi-complete LCS. Šmul'yan [4] shows the equivalence in B of (Š) and (wωC). Eberlein [1] was the first to break from any *countable* compactness condition to (wC).) Dieudonné uses Eberlein's device, used here in the proof of (Pc_0)$\Rightarrow$(E_r), to prove: *If A is compact and if E is a convex subset of R^A which consists entirely of continuous functions on A and which satisfies* (Š), *then the closure of E in R^A consists entirely of continuous functions on A.*

Pryce uses instead the Kaplansky-Whitley method to prove: *If S is a topological space and if $C(S)$ is the space of real-valued continuous functions on S with the relative topology from R^S, and if A is a* (cC) *subset of $C(S)$, then for each X in the closure of A in R^S and for each σ-compact subset B of S there is x_B in $C(S)$ and x_n in A such that $x_n(t)\to x(t)=x_B(t)$ for all t in B.*

(9) There are some useful conditions for weak compactness of sets in particular spaces. (a) If B is reflexive, a subset E of B is relatively weakly compact if and only if it is bounded: for example, for $1<p<\infty$ and arbitrary measure μ, $L^p(\mu)$ is reflexive. (b) For a bounded set M in $L^1(\mu)$ each of the following conditions are criteria for relative weak compactness of M (or for any of the many other conditions equivalent to (wC_r)): (i) The integrals $\int_E f\,d\mu$, f in M, are equi-absolutely continuous; that is, for each $\varepsilon>0$ there is $\delta(\varepsilon)>0$ such that for all f in M, $\int_E |f|\,d\mu<\varepsilon$ when $\mu(E)<\delta(\varepsilon)$. (ii) For each decreasing sequence (E_n) of measurable sets $\lim_n \int_{E_n} f\,d\mu=0$ uniformly for f in M. [Dunford and Schwartz, Ch. IV, § 8]. (c) Dunford and Schwartz, Ch. IV, § 6, also give a condition, quasi-equicontinuity, for weak compactness of a bounded set E in $C(S)$, S compact Hausdorff.

§ 3. Completely Continuous Linear Operators

We shall modify the classical definition of the property at hand, but the results of the preceding section give a number of possible alternative formulations.

Definition 1. A function T from one LTS into another is called *completely continuous* $[cc]$ if T carries each bounded set to a relatively compact set.

In this section we shall be concerned with linear operators completely continuous in the norm or the weak topology; these will be labelled cc and wcc operators, respectively.

(1) By Cor. 2,3 a linear operator T from one Banach space B into another B' is $cc[wcc]$ if and only if the image $T(U)$ of the unit ball U in B is norm [weak] relatively sequentially compact in B'. (b) A $[w\text{-}]cc$ linear operator defined on a Banach space is continuous.

(2) The family (CC) $[(wCC)]$ of all $cc[wcc]$ linear operators from B into B' is well behaved: (a) If such an operator is multiplied either before or after by a $[w\text{-}]$ continuous linear operator, the composed operator is of the same kind. (b) If (T_n) is a sequence of $[w\text{-}]cc$ linear operators and if $\|T_n - T\| \to 0$, then T is $[w\text{-}]cc$. (c) In the algebra $\mathfrak{L}(B)$ of all linear continuous operators from B into itself, $([w\text{-}]CC)$ is a norm-closed, two-sided ideal.

Let us now prove a deeper result due to Schauder [3].

Theorem 1. *A linear operator T from B into B' is cc if and only if T^* is cc.*

Proof. If T is cc and if for each n in $\omega, f_n \in U'^{\pi}$, the unit ball of B'^*, then let $g_n = T^* f_n$. To show that (g_n) has a norm-convergent subsequence observe that $T(U)$ is relatively compact, therefore totally bounded, so there exists a countable subset $\{x_n\}$ of U such that $\{Tx_n : n \in \omega\}$ is dense in $T(U)$. The set of f_n is equicontinuous and bounded on E, the closure of $T(U)$, because E is compact; by Ascoli's theorem there exists a sequence (f_{n_i}) uniformly convergent on E. Then $\|g_{n_i} - g_{n_j}\| = \sup\{|f_{n_i}(x') - f_{n_j}(x')| : x' \in T(U)\}$ tends to zero. By completeness of B^* the Cauchy sequence (g_{n_i}) has a limit in B^*. This proves that if T is cc, so is T^*.

If T^* is cc, by the preceding result T^{**} is cc; hence $T(U) = Q'^{-1}(T^{**}(Q(U)))$ is part of the relatively compact set $Q'^{-1}[Q'(B') \cap T^{**}(U^{\pi\pi})]$; hence T is cc.

(3) Examples are to be found in Banach, or Riesz and Nagy, where linear integral equations are discussed using the property that they are defined from completely continuous operators. In $l^p(\omega)$ coordinatewise multiplication by a fixed element of the space is cc.

(4) (a) If T is cc, then T carries every w-convergent sequence in B to a norm-convergent sequence in B'. (b) The converse is false; an example is to be found in $L^1(\mu)$. See Dunford and Pettis, or Theorem VI, 4, 5.

To prepare for the corresponding theorem for wcc operators requires

Lemma 1. *Let T be a linear and wcc operator from B into B'; then* (a) $T^{**}(U^{\pi\pi}) \subseteqq Q'$ *(norm closure of $T(U)$), and* (b) *if $(g_n\ n \in \Delta)$ is a net of elements of U'^{π}, then $w\text{-}\lim_{n\in\Delta} T^* g_n = 0$ if and only if $w^*\text{-}\lim_{n\in\Delta} T^* g_n = 0$.*

Proof. $T(U)$ is convex and w-relatively compact. By Cor. I. 6, 4 its norm-closure E is w-closed; by Eberlein's theorem 2,3 the w^*-closure of $Q'(T(U))$ is in $Q'(B')$, hence in $Q'(E)$. But $T^{**}(U^{\pi\pi})$ is, by Cor. II, 5, 5, part of the w^*-closure of $Q'(T(U))$; therefore it is part of $Q'(E)$.

Weak convergence of $T^* g_n$ always implies w^*-convergence. If $w^*\text{-}\lim_{n\in\Delta} T^* g_n = 0$, then for each x in U, $\lim_{n\in\Delta} T^* g_n(x) = \lim_{n\in\Delta} g_n(Tx) = 0$. Hence the net $(g_n, n \in \Delta)$ converges to zero on a norm-dense subset $T(U)$ of E. Hence $g_n(y) \to 0$ for every y in E; in particular, if $X \in U^{\pi\pi}$, then $X(T^* g_n) = T^{**} X(g_n) = g_n(Q'^{-1}\, T^{**} X)$ tends to zero. Hence $w\text{-}\lim_{n\in\Delta} T^* g_n = 0$.

Theorem 2 [Gantmaher]. *A linear transformation T from B into B' is wcc if and only if T^* is wcc.*

Proof. If T is *wcc*, choose a net $(g_n, n \in \Delta)$ in U'^{π}. By w^*-compactness of U'^{π}, (g_n) has a subnet $(h_m, m \in \Delta')$ such that (h_m) is w^*-convergent to some h in U'^{π}. By w^*-continuity of T^*, $T^* h_m - T^* h$ has w^*-limit 0 in B^*; by Lemma 1, $T^* h_m - T^* h$ has w-limit zero, or $T^* h$ is the w-limit of $T^* h_m$. Hence $T^*(U'^{\pi})$ is relatively w-compact, so T^* is *wcc*.

As in Theorem 1, T^* *wcc* implies T^{**} *wcc* implies T *wcc*.

Complete continuity can be used as an aid in proving compactness of a convex hull of a compact set. This proof is drawn from Phillips [2]; the result for w-ω-compactness is due to Kreĭn.

Lemma 2. *Let E be a relatively [weakly] sequentially compact subset of a Banach space B, and define T from $l^1(E)$ into B by $T\eta = \sum_{x\in E} \eta(x)\, x$ for each η in $l^1(E)$. Then T is a [w-] cc linear operator.*

Proof. In the appropriate topology, the closure of E is compact; hence it suffices to consider the lemma for compact E. E is bounded and B is complete so for each η the series for $T\eta$ is absolutely convergent in B to an element; indeed, if $E \subseteqq \{x; \|x\| \leqq K\}$, then for every finite subset σ of E, $\left\| \sum_{x\in\sigma} \eta(x) x \right\| \leqq K \sum_{x\in\sigma} |\eta(x)|$, and $T\eta = \lim_{\sigma\in\Sigma} T_\sigma \eta$, where each $T_\sigma \eta = \sum_{x\in\sigma} \eta(x)\, x$ and Σ is the set of finite subsets of E directed by $\supseteqq$. Then the T_σ are bounded linear operators on $l^1(E)$ with norm $\leqq K$, so $\|T\| \leqq K$.

To prove T is *cc*, we appeal to Theorem 1 or 2 for the appropriate case. T^* carries B^* into $l^1(E)^*$ and the natural isometry V of $l^1(E)^*$ onto $m(E)$ carries φ in B^* to a function $VT^*\varphi$ defined on E; it is easily verified that this function is just $\varphi_{|E}$; that is, φ with its domain of definition restricted to E, so VT^* carries U^π into a subset of a ball of $C(E)$.

Consider first the norm case. $VT^*(U^\pi)$ is bounded and equicontinuous on the compact set E. Ascoli's theorem asserts that this set is relatively compact in $C(E)$. Because V is an isometry, $T^*(U^\pi)$ is also relatively compact in $l^1(E)^*$; by Theorem 1 T is cc.

In the weak case E is w-compact but there is no w-equicontinuity so another device is needed. In case E has a countable w-dense subset $(x_n : n \in \omega)$, let $(f_i, i \in \omega)$ be any sequence in U^π. Then there is a subsequence (f_{i_k}) such that $\lim_{k\in\omega} f_{i_k}(x_n)$ exists for each n. Hence $\varphi(x) = \lim_{k\in\omega} f_{i_k}(x)$ exists for each x in L, the closed linear subset of B spanned by the x_n, and, in particular, for each x in E. $\varphi \in L^*$, so $\varphi_{|E}$ is in $C(E)$. To prove that in $C(E)$ the sequence $(VT^* f_{i_k}) = (f_{i_k|E})$ has the weak limit $\varphi_{|E}$, one possible method is to apply VI, 4, (3), because the sequence $(f_{i_k|E} - \varphi_{|E})$ is uniformly bounded and pointwise convergent to zero. This proves VT^* is wcc if E has a countable w-dense subset. But V is $w-w$ bicontinuous, so T^* is wcc; Theorem 2 says T is wcc.

In the general case, choose a sequence $(\eta_n, n \in \omega)$ from U_1, the unit ball of $l^1(E)$, and let $E_0 = \bigcup_{n\in\omega}\{x : \eta_n(x) \neq 0\}$, let E_1 be the w-closure of E_0, let $T_1 \eta = \sum_{x\in E_1} \eta(x)\, x$ if $\eta \in l^1(E_1)$. Then $T_1 \eta_n = T \eta_n$ for all n in ω, and, by the preceding case, T_1 is wcc. Hence the sequence $(T_1 \eta_n)$ has a w-convergent subsequence $(T_1 \eta_{n_k}) = (T \eta_{n_k})$ so T is also wcc.

(5) This lemma can also be proved without appeal to Eberlein's theorem or the representation for linear functionals on $C(E)$ by means of a lemma of Banach [p. 219] which characterizes, for countable E, those sequences in $m(E)$ which converge weakly to zero; see Phillips [2].

Theorem 3. *If E is a relatively [w-] compact subset of a Banach space B, so is $k(E)$, the convex hull of E. $K(E)$, the closed convex hull of E, and $E^\pi{}_\pi$ are [w-] compact.*

Proof. By Lemma 2 the function T defined there carries U_1 into a relatively [w-] compact set $T(U_1)$ such that $k(E) \subseteq T(U_1)$, which is dense in $E^\pi{}_\pi$.

(6) This is not the best proof available for the norm case, for the corresponding result is true in a quasi- (or even topologically) complete LCS L. If E is a totally bounded set in L, so is $E \cup -E = F$; $K(E) \subseteq K(F) = F^\pi{}_\pi$, which is totally bounded by III, 1, (1c). By topological completeness $K(E)$ is compact in L.

(7) If T is a [w-] cc linear operator from B into B, then the set of solutions of $Tx = x$ is finite dimensional [reflexive].

(8) (a) For each Banach space B, the space (CC) of all compact linear operators is a closed linear subspace and ideal in $L(B, B)$. (b) Let I

be the norm closure in $L(B,B)$ of the set of linear operators of finite rank; then I is the smallest closed ideal in $L(B,B)$; in particular, $I \subseteqq (CC)$. (c) Schatten proves that in Hilbert space $(CC)^*$ is the class of Hilbert-Schmidt operators and $(CC)^{**}$ is $L(H,H)$. (d) In Hilbert space and *in every other space with the approximation property* (Defn. IV, 3, 4), $I=(CC)$. (e) Enflo's example of a space without (AP), discussed at the end of IV, § 3, shows that this is not true even of all separable conjugate or reflexive spaces. [See IV, 3, (1b).]

An excellent discussion of completely continuous linear operators can be found in Riesz and Nagy.

§ 4. Reflexivity

We now know that all versions of w-compactness of a w-closed subset of a Banach space agree. We apply this to the problem of reflexivity of B; that is, to the investigation of properties of B equivalent to the condition $Q(B)=B^{**}$.

Theorem 1. *The following properties of a complete normed linear space B are equivalent:*

(A) *B is reflexive.*

(B) *U, the unit ball of B, is w-compact.*

(C) *B^* is reflexive.*

(D) *U^π is w-compact.*

Proof. (B) implies (A), for then $Q(U)$ is w^*-closed and also w^*-dense in $U^{\pi\pi}$ by Th. II, 5, 4. (A) implies that w- and w^*-topologies agree in B^*; with Theorem II, 5, 3 this gives (A) implies (D). Let $\mathfrak{i}$ be the identity operator in B; then (D) says that $\mathfrak{i}^*$ is wcc; Theorem 3,2 says that $\mathfrak{i}$ is also wcc. Since $\mathfrak{i}(U)=U$, (B) holds.

(1) To continue let $B^{(0)}=B$ and $B^{(n+1)}=B^{(n)*}$; let Q_n be the natural isometry from $B^{(n)}$ into $B^{(n+2)}$, let $\mathfrak{i}_n$ be the identity function in $B^{(n)}$, and let U_n be the unit ball in $B^{(n)}$. Then the relations "(A) implies (D) implies (B) implies (A)" proved above give the necessary induction step to prove that (C), as well as each of the following conditions, is equivalent to (A):

(E) One (or every) $B^{(n)}$ is reflexive.

(F) One (or every) U_n is w-compact.

(2) It is clear that all compactness conditions from § 2 (including $((w\omega C)$ and (J)) can be translated into conditions on the unit ball equivalent to reflexivity. From preservation of compactness come other equivalent conditions:

(G) B is isomorphic to a reflexive space.

(H) Every subspace of B is reflexive.

(I) Every factor space of B is reflexive.

(3) Corresponding to Cor. II, 4, 2 is another characteristic property of reflexive spaces:

(J) If $X \in B^{**}$ and if F is a set of elements of B^*, then there exists an x in B (even an x of the same norm as that of X) such that $X(f)=f(x)$ for all f in F.

(A) implies (J) trivially. The proof that (J) implies (A) uses $F=B^*$.

(4) Bourgin [2] gives a variant of total boundedness which is also equivalent to reflexivity:

(K) If $(f_n, n\in\Delta)$ is a net in B^* with 0 as a w^*-cluster point, then for each $\varepsilon>0$ and each n in Δ there is a finite set $n_1, \ldots, n_k$ in Δ with all $n_i \geqq n$ such that $U_0 \subseteqq \bigcup_{i\leqq k}\{x: |f_{n_i}(x)|<\varepsilon\}$.

To prove this observe that (K) implies:

(L) If the net $(f_n, n\in\Delta)$ is w^*-convergent to 0, then the conclusion of (K) holds. But by Cor. II, 4, 2, (L) implies:

(M) If $w^*\text{-}\lim_{n\in\Delta} f_n=0$, then $w\text{-}\lim_{n\in\Delta} f_n=0$. (M) and Th. II, 4, 3 imply (D); (B) implies (K).

(5) (a) For every B $(=B^{(0)})$, $Q_0^* Q_1 = \mathrm{i}_1$, in general, $Q_{n-1}^* Q_n = \mathrm{i}_n$. (b) Another condition equivalent to reflexivity of B is

(N) $Q_1 Q_0^* = \mathrm{i}_3$, or in general, $Q_{n+1} Q_n^* = \mathrm{i}_{n+3}$.

((N) implies that Q_1 is onto, which is (C). If $X=Q_0 x$, then for each Ξ in $B^{(3)}$, $Q_1 Q_0^* \Xi(X) = X(Q_0^* \Xi) = Q_0 x(Q_0^* \Xi) = (Q_0^* \Xi)(x) = \Xi(Q_0 x) = \Xi(X)$; hence $Q_1 Q_0^* \Xi = \Xi$ if B is reflexive.) (c) A similar condition is.

(O) $Q_0^{**}=Q_2$.

(Similar calculations give $\|Q_0^{**}X - Q_2 X\|$ = distance from X to $Q(B)$, so $Q_0^{**}=Q_2$ if and only if $Q(B)=B^{**}$. Dixmier [1] notes that if $\|X\|=1$, then the segment with ends at $Q_0^{**}X$ and $Q_2 X$ consists entirely of points of norm 1 in $B^{(4)}$; that is, *if B is not reflexive, then $B^{(4)}$ is not rotund.* Giles shows that $B^{(3)}$ is not smooth. [A simple proof uses James's property, (J) of § 2. Take a non-supporting functional f of norm one in B^* and an X in $B^{(2)}$ such that $\|X\| = X(f) = \|f\| = 1$; then X is not in $Q_0(B)$ so $Q_2 X \neq Q_0^{**} X$, but both support $U^{(3)}$ at $Q_1 f$.]

Theorem 2. *Each of the following conditions is equivalent to reflexivity of B:*

(P) *B is weakly quasi-complete.*

(Q) *B is bw-complete.*

Proof. (A)⇒(Q). When B is reflexive. Q^{-1} carries the bw^*-uniformity of B^{**} onto the bw-uniformity of B. But by Lemma II, 5, 3, B^{**} is bw^*-complete; hence B is bw-complete.

(Q)⇒(P). If B is bw-complete, if E is a bounded subset of B, and if (x_n) is a w-Cauchy net in E, then Ascoli's theorem implies that $x_n - x_m$ satisfies the conditions of Lemma II, 5, 2 (iii$_c$), so (x_n) is a bw-Cauchy net in B; the bw-limit x given by (Q) is also a w-limit, so $x \in w(\mathrm{E})$ and (P) holds.

(P)⇒(A). If $X \in B^{**}$, then by Cor. II, 4, 2 there is a net (x_n) in the ball of radius $\|X\|$ about 0 such that $X = w^*\text{-}\lim_{n \in \Delta} Q x_n$. Hence (x_n) is a bounded w-Cauchy net; by (P) it has a w-limit x in B, so $X(f) = \lim_{n \in \Delta} f(x_n) = f(x)$ for all f in B^*; that is, $X = Qx$.

(6) Many of these reflexivity criteria have quantitative variants, some of which are equivalent to reflexivity and some of which, are weaker. (a) For example, by Eberlein's Theorem III, 2, 3 two equivalent conditions are:

(B_ω) U is w-sequentially compact,

(B_0) U is w-countably compact.

(b) It follows from (B_0) that reflexivity of B is equivalent to

(H_0) Every separable subspace of B is reflexive.

(7) It is already known (Th. 2) that bw-completeness of B is equivalent to reflexivity and (Cor. II, 2, 3) that its countable variant

(P_ω) U is w-sequentially complete,

is necessary but insufficient for reflexivity. However, *if B^* is separable, (P_ω) implies reflexivity.*

(8) It is known that the following necessary conditions are not equivalent to reflexivity; by VI, 4, (4a), $m(S)$ satisfies (M_ω).

(I_0) Every separable factor space of B is reflexive.

(L_ω) If $(f_n, n \in \omega)$ is a sequence in B^* which is w^*-convergent to 0, then the conclusion of (K) holds.

(M_ω) If $(f_n, n \in \omega)$ is a sequence in B^* which is w^*-convergent to 0, then (f_n) is w-convergent to 0.

(a) (L_ω) and (M_ω) are equivalent properties of B. (b) If B has (M_ω), so has every factor space of B. (c) If a separable space has (M_ω), then it is reflexive. (d) (M_ω) implies (I_0). (e) If B satisfies (M_ω), then B^* satisfies (P_ω) [Day 9].

(9) Recall that a set E has James's property, § 2, (J), when every supporting hyperplane of E contains a point of E. (a) Each (wpC) set in each LCS has § 2 (J). (b) Klee [8] showed that a Banach space B is reflexive if and only if every symmetric convex body in B has § 2 (J), that is, if and only if every space isomorphic to B has unit ball with § 2 (J). (c) In a series of papers James [3,4,5] showed that § 2 (J) implies (wC) for w-closed sets in w-complete spaces. In [10] he returns to the reflexive case and gives a brief clear proof (§ 2, (7)) for separable spaces and a much shorter one than before for the general case.

(10) Some geometric properties of the unit ball U imply reflexivity of B; see VI, §§ 2 and 4. (a) Mil'man [8] showed that *every uniformly rotund space is reflexive.* (b) Šmul'yan [6] showed that *if the conjugate norm in B^* is Fréchet differentiable, then B is reflexive.* (c) Lovaglia showed that *if B^* is* (LUR), *then B is* (F); this with (b) shows that *if B^{**} is* (LUR), *then B is reflexive.* (d) James[6] showed that *uniformly non-square spaces are reflexive*, and that *not every reflexive space is isomorphic to a uniformly non-square space.* See VII, § 4 where there is a discussion of some work of Enflo [1] and of James [6 to 9] which describes a number of geometric properties of some very reflexive spaces, ending with the proof that such spaces are all isomorphic to uniformly rotund spaces. (See the last half of VII, § 4.)

(11) Much effort has been devoted to renorming spaces to get pleasanter analytic or geometric properties for the norm. This will be discussed in detail in VII, § 4, but here we note that the culmination of work by Lindenstrauss, Asplund, Troyansky, and others shows that: *Each reflexive space can be given an isomorphic norm such that both this norm and its conjugate are simultaneously Fréchet differentiable and locally uniformly rotund.* The basic step is the structure theorem, Th. 5.1 for non-separable reflexive spaces; the rest of the main steps are covered in Th. VII, 4, 1.

§ 5. Weak Compactness and Structure in Normed Spaces

Weakly compact subsets in normed spaces have special properties not shared by all compact Hausdorff spaces, for example, weak sequential compactness. In this section we discuss some other topological properties of such spaces and also give a structure theorem for normed spaces which possess a weakly compact fundamental set.

Lindenstrauss [1,2] for non-separable reflexive Banach space and Amir-Lindenstrauss for a larger class, the (WCG) spaces, construct and apply a "long sequence" of projections of norm one which gives much structural information about such a space, nearly as much as would the presence of a generalized basis. We state the theorem and give here a fairly full outline of the proof; applications to renorming for more rotundity or smoothness are given in VII, § 4, A.

(1) An *Eberlein compactum*, (EC), is a compact Hausdorff space which is homeomorphic to some w-compact subset of some Banach space. A *weakly compactly generated* (WCG) *space* is a Banach space with a weakly compact fundamental set.

(a) Each separable space, each reflexive space, and each $c_0(S)$ over each index set S is (WCG). (B has a norm-compact fundamental set if and only if B is separable, so that concept needs no separate name. The unit ball of a reflexive space is w-compact, by Thm. 4,1, (B). The unit ball of each $l^p(S)$, $1 \leq p < \infty$, is a w-compact fundamental set for $c_0(S)$.)

(b) This raises a problem connected to Lemma VII, 4,2. If B is (WCG), is there a reflexive B_0 and a linear continuous T mapping B_0 onto a dense subset of B? That the symmetric closed convex hull of a w-compact fundamental set need not itself be the unit ball of a reflexive space is illustrated by the unit ball of $l^1(S)$ in $c_0(S)$, but in that case there are other compact sets available; in particular the unit ball of the reflexive space $l^2(S)$ is a w-compact fundamental subset of $c_0(S)$. Asplund averaging (VII, 4, (4)) of the c_0 and l^1 norms in $l^1(S)$ seems to give the l^2 norm there. Is this averaging a general way to get a dense reflexive space in a (WCG) space

Theorem 1. *If B is a non-separable* (WCG) *space and μ is the first ordinal corresponding to the least cardinal number dens(B) of a dense subset of B, then there exists a transfinite sequence of projections P_α, $\omega \leq \alpha \leq \mu$, such that* (i) *$P_\mu$ is the identity.* (ii) *$\|P_\alpha\| = 1$ if $\omega \leq \alpha \leq \mu$.* (iii) *dens $(P_\alpha(B)) \leq$ the cardinal number of $\alpha <$ dens(B) if $\alpha < \mu$.* (iv) *If $\omega \leq \alpha \leq \beta \leq \mu$, then $P_\alpha P_\beta = P_\beta P_\alpha = P_\alpha$.* (v) *If S_μ is the space of all ordinals $\leq \mu$ with the order topology, then the transfinite sequence (P_α) maps B into $C_\mu(B)$, the space of continuous functions from S_μ to B; that is, if β is a limit ordinal and $\omega < \beta \leq \mu$, then $\lim_{\alpha<\beta} \|P_\alpha x - P_\beta x\| = 0$ for each x in B.* (vi) *Let $\tau_\alpha = P_{\alpha+1} - P_\alpha$, $\omega < \alpha < \mu$, and for each x let $\Lambda(x) = \{\alpha : \tau_\alpha(x) \neq 0\}$. Then each $\Lambda(x)$ is countable, and if $\beta = \sup(\Lambda(x)+1)$, then $x = P_\beta(x) = P_\omega x + \sum_{\alpha \in \Lambda(x)} \tau_\alpha x$.* (Note that $\Lambda(x)$ is well ordered and countable but is not necessarily isomorphic to the ordinal ω.) (vii) *x belongs to Y_x, the closed linear hull of $P_\omega B + \sum_{\alpha \in \Lambda(x)} \tau_\alpha(B)$.*

Long sketch of proof. Take a symmetric, convex, w-compact, fundamental set K inside U, the unit vall of B, and let N be the linear hull of K; let $\|x\|_K$ be the Minkowski functional of K. Then $\|x\| \leqq \|x\|_K$ for each x in N.

(2) Let A be a fixed finite-dimensional subspace of N and E a fixed finite-dimensional subspace of $N^* (= B^*)$.

(a) For each $\varepsilon > 0$ and each n and q in ω there is a finite-dimensional subspace C of N such that for each subspace $L \supseteq A$ with $\dim(L/A) = n$ and $\|x\|_K \leqq q\|x\|$ for all x in L there is an ε-isomorphism T of L into C; that is, $Ta = a$ if $a \in A$, $|\,\|Tx\| - \|x\|\,| < \varepsilon\|x\|$ and $|\,\|Tx\|_K - \|x\|_K| < \varepsilon\|x\|$ if $x \in L$, and $|f(Tx) - f(x)| < \varepsilon\|f\| \cdot \|x\|$ if $x \in L$ and $f \in E$. (The proof uses the compact set of possible norms and K-norms under $q\|\ \|$ and linear functionals of norm ≤ 1 definable on the product of A with an

n-dimensional l^1 space, we illustrate with $n=1$; the general case can be done similarly or by induction. In $A\times R$ every unit ball extending that of A and lying between the hyperplanes $(0,\pm 1)+A$ must be contained in the "diablo" $K_1\cup -K_1$, where K_1 is the convex hull of $(0,-1)$ and $(0,1)+2(A\cap U)$ (provided that coordinates are first selected so that $(0,\pm 1)$ are points of contact of the ball with the hyperplanes). The set of such norms is equicontinuous, hence compact. The extended K-norms are similarly restricted to be equicontinuous, and, since $E\cap U^{\pi}$ is finite dimensional, so are the sets of extensions to $A\times R$ because each extension is determined by its value at $(0,1)$. Now for each allowed L choose $\pm b_0$ in L to be points of contact of $U\cap L$ with the supporting hyperplanes parallel to A. Mapping $a+tb_0$ to (a,t) in $A\times R$ gives copies of $\| \ \|$ and $\| \ \|_K$ in $A\times R$ and of $(E\cap U^{\pi})(b_0)$ to attach at $(0,1)$. These are a subset of the compact set we had earlier so have an approximating ε-net coming from certain subspaces $L_1,\dots,L_m$. Let C be the linear hull of these L_i, $i\leq m$.)

(b) There is a separable closed linear subspace D of N such that whenever $\dim(L/A)\leq n$ there is an ε-isomorphism (as in (a)) of L into D. (Take C_q to be the C of (a) for each q in ω and let D be the closed linear hull of these C_q.)

(c) To go on requires the w-compactness of K; then $(2K)^K$ is a compact set of functions in the product of the weak topologies. Now there is a separable closed subspace C of N and a linear mapping T of N into C such that $Ta=a$ if $a\in A$, $T^*f=f$ if $f\in E$, $\|T\|=1$, and $\|T\|_K=1$. (For each n in ω set $\varepsilon=1/n$ and let D_n be the appropriate space from (b) for those L with $\dim(L/A)\leq n$, and take the closed linear hull of the D_n to be C. For each $L\supseteq A$ for which $\dim(L/A)\leq n$ define T_L from L into C with the ε-isomorphism properties; extend T_L by defining $T_L=0$ for all x not in L (so T_L is now homogeneous but not additive). Each T_L maps K into $2K$; direct the family $\{L\}$ by $\supseteq$ and take a convergent subnet in $(2K)^K$ to define T from K into C; extend it by homogeneity.)

(3) (a) After all this work the first major goal is reached.

Lemma 1. *For each separable subspace A of N, a* (WCG) *space, and each w^*-separable subspace E of N^* there is a projection P from N into itself such that* (i) *$P(N)$ is separable.* (ii) *$P(N)\supseteq A$.* (iii) *$P^*(N^*)\supseteq E$.* (iv) *$\|P\|=1$ and $\|P\|_K=1$.*

(Take $(x_i, i\in\omega)$ to be dense in A and $(f_i, i\in\omega)$ to be w^*-dense in E; let A_1 be the linear hull of x_1 and E_1 that of f_1; use (10c) to define T_1 and let $(x_{1i}, i\in\omega)$ be a dense subset of $T_1(N)$. If A_k, E_k, T_k, and (x_{ki}) are known for $k\leq n$ and all i in ω, let A_{n+1} be the linear hull of $\{x_i, x_{ki}: k\leqq n, i\leqq n+1\}$, E_{n+1} be the linear hull of $\{f_i: i\leqq n+1\}$, choose T_{n+1} by (10c) for these spaces and let $(x_{n+1\,i}: i\in\omega)$ be a dense subset of $T_{n+1}(N)$. By weak compactness of K, some subnet of $(T_k, k \in \omega)$ converges weakly to a P, which can be shown to have the properties listed.)

(b) *Same conclusion with* dens(A) *and* w^*-dens$(E) \leq m <$ dens(N). (The case $m = \aleph_0$ is the lemma. If $m > \aleph_0$ assume that the result is known for cardinals $n < m$ and let γ be the first ordinal of cardinal m. Let $(x_\alpha, \alpha < \gamma)$ $[(f_\alpha, \alpha < \gamma)]$ be dense [w^*-dense] in A [in E] and let $A_\omega [E_\omega]$ be spanned by $(x_i, i \in \omega)$ $[(f_i, i \in \omega)]$. By (a) there is a P_ω such that $P_\omega(N)$ is separable, $P_\omega(N) \supseteqq A$, and $P^*_\omega(N^*) \supseteqq E$. If P_α are already defined for $\omega \leq \alpha \leq \beta$, define $P_{\beta+1}$ so that $P_{\beta+1}(N) \supseteqq P_\beta(N)$ and $x_{\beta+1}$, $P_{\beta+1}{}^*(N^*) \supseteqq P^*_\beta(N^*)$ and $f_{\beta+1}$. If β is a limit ordinal and P_α are defined for $\omega \leq \alpha < \beta$, there is a w-convergent subnet of $(P_\alpha, \alpha < \beta)$; call its limit P_β. This defines P_β for all $\omega \leq \beta \leq \gamma$; in particular P_γ is a projection with the properties required.)

(c) The theorem is nearly here. If μ is the first ordinal of cardinal dens$(N) > \aleph_0$, construct P_ω from (a). If for $\omega \leq \beta \leq \mu$ the projections $P_\omega, \ldots, P_\beta$ are defined and commutative, construct $P_{\beta+1}$ for all $\alpha \leq \beta$ from (b) so that $P_{\beta+1}(N) \supseteqq P_\beta(N)$ and $x_{\beta+1}$, $P_{\beta+1}{}^*(N^*) \supseteqq P^*_\beta(N^*)$ and $f_{\beta+1}$. Then $P_{\beta+1} P_\alpha = P_\alpha = P_\alpha P_{\beta+1}$. If P_α are defined and commute for all $\alpha <$ some limit ordinal $\beta \leq \mu$, let $P_\beta = w$-limit of some w-convergent subnet of P_α, $\alpha < \beta$, so P_β also commutes with the earlier ones. These projections have the necessary properties in that their extensions to B using uniform continuity have the properties listed in Theorem 1. Note that (v) is the hardest to prove. By the definition for a limit ordinal β, $w\text{-}\lim_{\alpha_i} P_{\alpha_i} x = P_\beta x$, so $\bigcup_{\alpha < \beta} P_\alpha(B)$ is w-dense, hence norm dense, in $P_\beta(B)$. Hence for each $\varepsilon > 0$ there is $\gamma < \beta$ and x_γ in $P_\gamma(B)$ such that $\|x_\gamma - x\| < \varepsilon$; then for $\gamma \leq \alpha < \beta$, $\|x_\gamma - P_\alpha x\| < \varepsilon$ so $\|P_\alpha x - x\| < 2\varepsilon$; that is, $\|P_\alpha x - x\| \to 0$. From this it readily follows that $\|\tau_\alpha x\|$, $\alpha < \mu$ exceeds each positive ε on at most a finite set of α. This will yield (vi) and (vii).

(4) Troyanski [2] refines the Amir-Lindenstrauss decomposition until $\tau_\alpha(B) = (P_{\alpha+1} - P_\alpha)(B)$ is *separable* for each $\alpha < \mu$, but at the cost of losing the uniform bound on the norm; he applied this result to renorming of (WCG) spaces to get local uniform rotundity; see Lemma VII, 4, 3. [Let B be a (WCG) space and let μ be, as in Thm. 1, the first ordinal of cardinal dens(B). Then there is a transfinite sequence P_α, $\omega \leq \alpha \leq \mu$, of projections in B satisfying the following conditions: (i) P_μ is the identity (ii) $\|P_\alpha\| < \infty$ for each α. (iii) Each $\tau_\alpha(B)$, $\omega \leq \alpha < \mu$, is separable. (iv) If $\omega \leq \alpha \leq \beta \leq \mu$, then $P_\alpha P_\beta = P_\beta P_\alpha = P_\alpha$. (v) If for x in B and $\varepsilon > 0$, we define $\Lambda(x, \varepsilon) = \{\alpha : \|\tau_\alpha x\| > \varepsilon(\|P_\alpha\| + \|P_{\alpha+1}\|)$, then $\Lambda(x, \varepsilon)$ is a finite set. (vi) If $\Lambda(x) = \bigcup_{\varepsilon > 0} \Lambda(x, \varepsilon)$ and Y_x is the closed linear hull of $\|P_\omega x\| P_\omega(B) + \bigcup_{\alpha \in \Lambda(x)} \tau_\alpha(B)$, then $x \in Y_x$. (The proof works by transfinite induction starting from the Amir-Lindenstrauss decomposition Q_α, $\alpha \leq \mu$. This has $(Q_{\alpha+1} - Q_\alpha)(B)$ separable if μ is the first uncountable ordinal. For larger card μ, use the induction hypothesis to get a transfinite sequence $Q_{\alpha\beta}$, $\omega \leq \beta \leq \mu_\alpha < \mu$

satisfying the conditions above in $B_\alpha=(Q_{\alpha+1}-Q_\alpha)(B)$. Then define $P_{\alpha\beta}=Q_\alpha+Q_{\alpha\beta}(Q_{\alpha+1}-Q_\alpha)$. (The (α,β) pairs will then be lexicographically ordered and can be relabelled with ordinals α with $\omega\leq\alpha<\mu$.) It is a dull exercise to show these P_α have the properties claimed.)]

The next results come from Amir-Lindenstrauss.

Theorem 2. (a) *For each* (WCG) *space N there is a one-to-one linear continuous T from N into some $c_0(S)$. Hence every Eberlein compact [every w-compact subset of every normed space] is homeomorphic [affinely] to a w-compact subset of some $c_0(S)$.*

Proof. Let μ be the smallest ordinal of cardinal dens (N). If $\mu=\omega$, then N can be mapped into $c_0(\omega)$ by using a total sequence $(f_i, i\in\omega)$ of elements of norm one and setting $Tx=(f_i(x)/2^i, i\in\omega)$ as in Lemma VII, 4,1. If $\mu>\omega$, use the long sequence of Theorem 1 to get projections P_α, $\alpha<\mu$, with smaller density of their ranges, use an induction hypothesis that each $\tau_\alpha(B)$, $\alpha<\mu$, can be squeezed into some $c_0(S_\alpha)$ by means of a T_α of norm one. Let S be the disjoint union of the S_α and define T on N by $[Tx](s_\alpha)=[T_\alpha(P_{\alpha+1}-P_\alpha)x](s_\alpha)$, for all $\alpha<\mu$ and s_α in S_α.

If K_1 is any w-compact set in any normed N apply this result to K, the closed convex hull of $K_1\cup -K_1$ in B, the completion of N. Then K is w-compact also, by Thm. 3,3, and T is linear, w-continuous, and one-to-one into $c_0(S)$, so T is an affine homeomorphism of K_1 onto a w-compact subset of $c_0(S)$.

Corollary 1. *If K is an* (EC), *then $C(K)$ is* (WCG).

Proof. Represent K as a w-compact subset of the ball $U/2$ in some $c_0(S)$. For each finite sequence $\sigma=(s_1,\ldots,s_n)$ of (not necessarily distinct) elements of S construct f_σ in $C(K)$ by: For each k in K, $f_\sigma(k)=\prod_{i\leq n} k(s_i)$. Add to these f_σ the constant functions 0 and 1 to get a set $M\subseteqq C(K)$; then every not-ultimately-constant sequence in M has a subsequence weakly convergent to zero, so M is (wωC). The Stone-Weierstrass Theorem, VI, 2,1, says that the linear hull of M is dense in $C(K)$.

Corollary 2. *If B is* (WCG), *then there is a one-to-one, linear, bounded, T from B^* into some $c_0(S)$ which is continuous from the w*-topology of B^* to the w-topology of $c_0(S)$.*

Proof. If K is w-compact and fundamental in B, then Th. 3.3 says that K may be assumed convex. Map B^* into $A(K)\subseteqq C(K)$ by the restriction, ρf, of each f in B^* to K. V, 2, (2b) asserts that ρ is w^*-continuous from B^* into $C(K)$; Corollary 1 completes the proof.

Corollary 3. *If B is* (WCG), *then the unit ball of B^* in its w*-topology is an Eberlein compactum.*

Corollary 4. *For a compact Hausdorff space K the following conditions are equivalent:* (i) *K is an* (EC). (ii) *$C(K)$ is* (WCG). (iii) (U^π, w^*) *in $C(K)^*$ is an* (EC).

(5) Corson-Lindenstrauss showed that w-compact subsets of $c_0(S)$ have special properties which the Amir-Lindenstrauss result now extends to all w-compact subsets of all B-spaces. (a) Every w-compact convex K in a Banach space is the norm-closed convex hull of its set of exposed points (V, 1, (7)). (Corson-Lindenstrauss for $c_0(S)$; then if f exposes Tx in TK in $c_0(S)$, T^*f exposes x in K. The image by T^{-1} of a w-dense set in TK is w-dense in K.) (b) If B is (WCG), then by a similar argument each w^*-compact convex subset K of B^*, for example, the unit ball U^π, is the w^*-closure of its w^*-exposed points. (Compare this with the Krein-Mil'man Theorem V, 1, (4), and with Lindenstrauss's result V, 1, (8a) which with Th. VII, 4, 1 and the isomorphism invariance of strong exposure implies that *every* (wC) *convex set in a normed space is the closed convex hull of its set of strongly exposed points.*) (c) The G_δ points are a dense subset of any Eberlein compactum. (Look at the w^*-exposed points of the positive face of the unit ball in $C(K)^*$ which are, by Thm. V, 1, 3, w^*-dense in the set of extreme points, which is, by Thm. V, 3, 1, (ii), homeomorphic to K.

Chapter IV. Unconditional Convergence and Bases

§ 1. Series and Unconditional Convergence

We shall be interested in applications mainly in weak and norm topologies of a Banach space, but first we describe several possible forms of convergence of a series $\sum_{i\in\omega} x_i$ of elements of an LCS L.

(A) $\sum_{i\in\omega} x_i$ is *ordered convergent* (briefly, *convergent*) to an element x of L if $\lim_{n\in\omega} \sum_{i\leq n} x_i = x$.

(B) $\sum_{i\in\omega} x_i$ is *reordered convergent* to x if for every permutation p of the integers the series $\sum_{i\in\omega} x_{p(i)}$ is convergent to x. This is the classical formulation of unconditional convergence.

(C) $\sum_{i\in\omega} x_i$ is *unordered convergent* to x if, letting Σ be the system of finite subsets σ of ω, directed by $\supseteqq$, $\lim_{\sigma\in\Sigma} \sum_{i\in\sigma} x_i = x$.

(D) $\sum_{i\in\omega} x_i$ is *subseries convergent* if for every increasing sequence $(n_i, i\in\omega)$ of integers the series $\sum_{i\in\omega} x_{n_i}$ is convergent (to some element of L).

(E) $\sum_{i\in\omega} x_i$ is *bounded-multiplier convergent* if for each bounded real sequence $(a_i, i\in\omega)$ the series $\sum_{i\in\omega} a_i x_i$ is convergent to an element of L.

(F) $\sum_{i\in\omega} x_i$ is *absolutely convergent* if for each neighborhood U of 0 in L the series of non-negative numbers $\sum_{i\in\omega} p_U(x_i)$ is convergent and the given series is convergent; here p_U is the Minkowski functional of U.

If the Cauchy condition is substituted for convergence, six related conditions $(A_c), \ldots, (F_c)$ are obtained.

(1) (a) $(F_c)\Rightarrow(E_c)\Leftrightarrow(D_c)\Leftrightarrow(C_c)\Leftrightarrow(B_c)\Rightarrow(A_c)$. (b) $(E)\Rightarrow(D)\Rightarrow(C)\Leftrightarrow(B)\Rightarrow(A)$; $(F)\Rightarrow(B)$. (c) $(X)\Rightarrow(X_c)$ for $X=A, \ldots, F$. (d) Hence when L is sequentially complete, $(X_c)\Leftrightarrow(X)$ and $(F)\Rightarrow(E)\Leftrightarrow(D)\Leftrightarrow(C)\Leftrightarrow(B)\Rightarrow(A)$.

[To illustrate this, let us prove (D_c) implies (C_c). Suppose that $\sum_{i\in\omega} x_i$ is not unordered Cauchy; then there is a neighborhood U of 0 and a sequence (σ_k) of pairwise disjoint finite sets of integers such that for each k in ω, $\sum_{i\in\sigma_k} x_i$ is not in U. Let (τ_j) be a subsequence of (σ_k) such that for each j in ω, $\inf \tau_{j+1} > \sup \tau_j$; then enumerate the union of the (τ_j) in order of size as a sequence (i_n). Then $\sum_{n\in\omega} x_{i_n}$ is not Cauchy.] (See Hildebrandt [2], Orlicz [2].)

(2) (a) Riemann's theorem [de la Vallee-Poussin, p. 419] asserts that *if* L *is finite-dimensional, then* (B)⇒(F); that is, *reordered and absolute convergence are equivalent properties of a series.* (b) The example $x_i = \delta^i/i$ in $c_0(\omega)$ shows that (B) need not imply (F) in all normed spaces.

(3) (F) is also equivalent in a normed space N to a kind of "bounded-multiplier" convergence. (a) If x, y are any two elements of norm 1 and L is the linear hull of x, y, the linear operator T' in L which interchanges x with y is of norm $\leqq 3$ in L. (b) Every one-dimensional subspace of N has (by the Hahn-Banach theorem) a projection on it of norm 1; hence there is a projection P of N on L of norm $\leqq 2$. (c) Hence $T = T'P + (I - P)$ is an isomorphism of norm $\leqq 9$ in N which interchanges x and y. (d) Aligning each $T_n x_n$, $n > 1$, with x_1 shows that in a Banach space B the condition (F) is equivalent to: (E′) for every bounded sequence $(T_i, i\in\omega)$ of linear operators on B, $\sum_{i\in\omega} T_i x_i$ is convergent to an element of B. In a normed space N, (F_c) is equivalent to a corresponding condition (E'_c).

(4) (a) If the w-topology is used in an LCS L, then the absolute and the unordered convergent [Cauchy] series are the same. (b) If $\sum_{i\in\omega} x_i$ is a series in a Banach space B, then $K = \sup\left\{\left\|\sum_{i\in\sigma} x_i\right\| : \sigma\in\Sigma\right\} < \infty$ if and only if $K' = \sup\left\{\left\|\sum_{i\in\sigma} a_i x_i\right\| : |a_i| \leqq 1 \text{ for all } i \text{ and } \sigma\in\Sigma\right\} < \infty$ (indeed, $K \leqq K' \leqq 2K$), and if and only if $\sum_{i\in\omega} x_i$ is weakly unordered Cauchy. (c) If B is a Banach space and if $\sum_{i\in\omega} x_i$ is weakly unordered Cauchy, then F, defined by $Ff = (f(x_i), i\in\omega)$, is a bounded linear operator from B^* into $l^1(\omega)$. (d) If in a Banach space B, $\sum_{i\in\omega} x_i$ is weakly subseries convergent, then it is weakly bounded-multiplier convergent and T, defined for each $a = (a_i, i\in\omega)$ in $m(\omega)$ by $Ta = w - \sum_{i\in\omega} a_i x_i$, is a bounded linear operator from $m(\omega)$ into B, and the F of (c) is $Q^* T^*$. (e) If B is a w-sequentially complete Banach space, then (w- or norm-) boundedness of the set of all partial sums $\sum_{i\in\omega} x_i$ is equivalent to unordered (or to absolute) w-convergence of $\sum_{i\in\omega} x_i$.

Theorem 1. *If in the w-topology of a Banach space B the series* $\sum_{i\in\omega} x_i$ *is subseries convergent, then it is subseries convergent in the norm topology of B.*

[Remark. Orlicz [1] proved this for w-ω-complete spaces; the same proof works in general, see Pettis [2].]

Proof. Let L be the smallest closed linear subset of B containing the x_i; then by (4d), Ta, the weak unordered sum $\sum_{i\in\omega} a_i x_i$, defines a bounded linear function T from $m(\omega)$ into L, so Cor. II,2,1 applies.

Take g_i in L^* such that $\|g_i\|=1$ and $g_i(x_i)=\|x_i\|$; then (because L is separable) there is a w^*-convergent subsequence $(h_j)=(g_{i_j})$ with limit h. $\sum_{i\in\omega} h(x_i)$ is convergent so $h(x_i)\to 0$. If $z_j=x_{i_j}$ and $f_j=h_j-h$, we have $w^*\text{-}\lim_{j\in\omega} f_j=0$ and $\lim_{j\in\omega}(f_j(z_j)-\|z_j\|)=0$. By (4d) and Cor. II,2,1, $\sum_{i\in\omega}|f_j(x_i)|\geqq|f_j(z_j)|\to 0$, so $\|z_j\|\to 0$.

This proves that if $\sum_{i\in\omega} x_i$ is weakly subseries convergent, then some subseries has elements tending in norm to zero. But if $\sum_{i\in\omega} x_i$ is not subseries convergent in norm, then, by completeness of B, it is not unordered Cauchy. Hence there exists $\varepsilon>0$ and disjoint finite sets τ_j (as in the proof after (2)) such that setting $y_j=\sum_{i\in\tau_j} x_i$ yields $\|y_i\|>\varepsilon$ for all j in ω. *But* $\sum_{j\in\omega} y_j$ is also weakly subseries convergent, so has a subseries with terms norm-convergent to 0; this contradiction proves that $\sum_{i\in\omega} x_i$ is unordered convergent and subseries convergent.

The main purpose of this section is to show that (2b) is not an accidental property of $c_0(\omega)$ but is valid in all infinite-dimensional Banach spaces. Similar examples were known in all the familiar spaces, but the normed case was finally settled by Dvoretsky and Rogers in the following theorem using, for example, $c_n=1/n$.

Theorem 2. *Let B be an infinite-dimensional Banach space and let* $(c_n, n\in\omega)$ *be any sequence of positive numbers such that* $\sum_{n\in\omega} c_n^2<\infty$; *then there exists in B an unordered convergent sequence* $(x_n, n\in\omega)$ *such that* $\|x_n\|=c_n$ *for every n in* ω.

First we prove a geometrical lemma about n-dimensional Euclidean spaces and symmetric convex bodies there.

Lemma 1. *Let B be an n-dimensional normed space; then there exist points* $x_1,\ldots,x_n$ *of norm one in B such that for each* $i\leqq n$ *and all real* $t_1,\ldots,t_i$

(a) $$\Big\|\sum_{j\leqq i} t_j x_j\Big\|\leqq\left[1+\left(i\,\frac{i-1}{n}\right)^{\frac{1}{2}}\right]\Big(\sum_{j\leqq i} t_j^2\Big)^{\frac{1}{2}}.$$

Proof. Inscribe in C, the unit ball of B, the ellipsoid E of maximum volume. [More precisely, if $u_1, \ldots, u_n$ is any vector basis in B, an ellipsoid E' is the set where some positive-definite quadratic form in the components, $Q(t_1, \ldots, t_n) = \sum_{i,j \leqq n} a_{ij} t_i t_j$, is $\leqq 1$; then the volume of E' is the square root of the reciprocal of determinant $|a_{ij}|_{i,j \leqq n}$, and $E' \subseteqq C$ means $\left\| \sum_{i \leqq n} t_i u_i \right\| \leqq 1$ if $Q(t_1, \ldots, t_n) \leqq 1$.] By a linear transformation of B we may turn E into the Euclidean ball whose coordinates satisfy $\sum_{i \leqq n} t_i^2 \leqq 1$. Using subscripts to distinguish the norm with unit ball E from that with unit ball C, we want now to search for x_i with $\|x_i\|_E = \|x_i\|_C = 1$ and the x_i approximately orthogonal. By induction on i we shall find an orthonormal basis $u_i, i \leqq n$, in the Euclidean space determined by E and points $x_i, i \leqq n$, with $\|x_i\|_E = \|x_i\|_C = 1$, such that

(i) $x_i = \sum_{j \leqq i} a_{ij} u_j$, and all $a_{ij} \geqq 0$, and

(ii) $a_{i1}^2 + \cdots + a_{ii-1}^2 \leqq (i-1)/n$.

To begin the proof take $u_1 = x_1$ to be any point of contact of the surfaces of C and E, and, for the moment let $u_2, \ldots, u_n$, be any vectors completing an orthonormal basis in E.

Suppose that $x_1, \ldots, x_i$ and $u_1, \ldots, u_i$, $1 \leqq i < n$, have been found to satisfy (i) and (ii) for all $j \leqq i$; fill out an orthonormal basis with any suitable $u_{i+1}, \ldots, u_n$ and consider for $\varepsilon > 0$ the "spheroid" E_ε of points whose coordinates in this basis satisfy

$$\text{(b)} \qquad (1+\varepsilon)^{n-i}(\alpha_1^2 + \cdots + \alpha_i^2) + (1+\varepsilon+\varepsilon^2)^{-i}(\alpha_{i+1}^2 + \cdots + \alpha_n^2) \leqq 1.$$

The volume of E_ε is easily calculated to be greater than that of E, so there is a point in E_ε outside C, and, by coming toward 0 along a ray, there is in E_ε a point p_ε of C-norm 1. Then p_ε is not inside E, so, if $\alpha_1, \ldots, \alpha_n$ are the coordinates of p_ε, $\sum_{j \leqq n} \alpha_j^2 \geqq 1$. Subtracting this from (b) gives

$$\text{(c)} \quad [(1+\varepsilon)^{n-i} - 1][\alpha_1^2 + \cdots + \alpha_i^2] + [(1+\varepsilon+\varepsilon^2)^{-i} - 1](\alpha_{i+1}^2 + \cdots + \alpha_n^2) \leqq 0.$$

By compactness, there is a subsequence of ε's tending to 0 such that the corresponding p_ε converge to some point x_{i+1} common to the surfaces of C and E. Dividing (c) by ε and taking the limit such a sequence gives, if x_{i+1} has coordinates $a_1, \ldots, a_n$,

$$\text{(d)} \qquad (n-i)(a_1^2 + \cdots + a_i^2) + (-i)(a_{i+1}^2 + \cdots + a_n^2) \leqq 0.$$

Choosing u_{i+1} orthogonal to $u_1, \ldots, u_i$ in the space spanned by these and x_{i+1} and completing this sequence to a new orthonormal basis gives a representation for x_{i+1} which can now be seen to satisfy the conditions (i) and (ii) for $i+1$. This induction process defines x_i and u_i for all $i \leqq n$.

(a) follows from the other conditions. (ii) implies that $\|x_i - u_i\|_E^2 = \left((1-a_{ii})^2 + \sum_{j<i} a_{ij}^2\right) \leqq 2(i-1)/n$. Since E is inside C, $\|x\|_C \leqq \|x\|_E$ for every x, so

$$\Big\|\sum_{j\leqq i} t_j x_i\Big\|_C \leqq \Big\|\sum_{j\leqq i} t_j u_j\Big\|_C + \Big\|\sum_{j\leqq i} t_j(x_j - u_j)\Big\|_C$$

$$\leqq \Big\|\sum_{j\leqq i} t_j u_j\Big\|_E + \sum_{j\leqq i} |t_j| \left(2\frac{j-1}{n}\right)^{\frac{1}{2}}.$$

Use of Schwarz's inequality shows that this

$$\leqq \left(\sum_{j\leqq i} t_j^2\right)^{\frac{1}{2}} + \left(\sum_{j\leqq i} t_j^2\right)^{\frac{1}{2}} \left(\sum_{j\leqq i} 2\frac{j-1}{n}\right)^{\frac{1}{2}}$$

$$= \left[1 + \left(i\frac{i-1}{n}\right)^{\frac{1}{2}}\right] \left(\sum_{j\leqq i} t_j^2\right)^{\frac{1}{2}}.$$

This is (a), from which we see that the constant is small if n is large compared to i. However, for the present purpose the following consequence of (a) suffices.

Lemma 2. *If $n \geqq i\,(i-1)$ then in every n-dimensional normed space B there exist i unit vectors $x_1, \dots, x_i$, such that* $\Big\|\sum_{j\leqq i} t_j x_j\Big\| \leqq 2 \Big\|\sum_{j\leqq i} t_j^2\Big\|^{1/2}$.

Proof of Theorem 2. Take (c_k) such that $\sum_{k\in\omega} c_k^2 < \infty$. Then group the terms in this sum in blocks $\sigma_j = \{k\colon N_j < k \leqq N_{j+1}\}$ so that $\sum_{j\in\omega} \left(\sum_{k\in\sigma_j} c_k^2\right)^{1/2}$ is convergent. In B find a sequence of linearly independent, finite-dimensional subspaces B_j each of such high finite dimension that in B_j vectors $A_{N_j+1}, \dots, A_{N_{j+1}}$ satisfying Lemma 2 can be constructed. Let $x_k = c_k A_k$; then $\|x_k\| = c_k$ and we need only test by the unordered Cauchy criterion to see whether $\sum_{k\in\omega} x_k$ is an unordered convergent series.

If σ is a finite set of integers, then

$$\Big\|\sum_{k\in\sigma} x_k\Big\| \leqq \sum_{j\in\omega} \Big\|\sum_{k\in\sigma\wedge\sigma_j} x_k\Big\| \leqq \sum_{j\in\omega} 2\left(\sum_{k\in\sigma\wedge\sigma_j} c_k^2\right)^{\frac{1}{2}}.$$

This can be made arbitrarily small by keeping σ disjoint from enough of the initial blocks of terms in the series, so $\sum_{k\in\omega} x_k$ is unordered Cauchy; by completeness of B it is unordered convergent.

If c_k are properly chosen $\sum_{k\in\omega} c_k = +\infty$; this gives an example of a series in B which is subseries convergent but not absolutely convergent.

(5) A partial converse of the Dvoretsky-Rogers theorem, known for many years, has been improved recently. (a) Orlicz [1] showed that if $(x_i, i\in\omega)$ is a sequence in $l^p(\omega)$ and if $\sum_{i\in\omega} x_i$ is unconditionally con-

vergent, then $\sum_{i\in\omega} \|x_i\|^q < \infty$ where $q=2$ if $1 \leqq p \leqq 2$, and $q=p$ if $p \geqq 2$. (b) Kadec [3] showed that if B is uniformly rotund (Chapter VII, 2) and if $\delta(\varepsilon)$ is the modulus of rotundity in B and if $\sum_{i\in\omega} x_i$ is unconditionally convergent in B, then $\sum_{i\in\omega} \delta(\|x_i\|) < \infty$. (c) For $p \geqq 2$, Clarkson's estimate of $\delta(\varepsilon)$ in $l^p(\omega)$ is precise and gives Orlicz's result from Kadec's; for $1 < p \leqq 2$, Kadec gives a closer estimate of $\delta(\varepsilon)$ (also found a little earlier by Hanner) which yields Orlicz's result for $p > 1$.

(6) Unordered and absolute convergence can be defined for series $\sum_{s\in S} x_s$ defined on an arbitrary index set S, just as in II, § 2; see Hildebrandt [1]. (a) In any LCS an absolutely Cauchy series is unordered Cauchy, but the converse need not be true. (b) If S and T are index sets, for every t in T, L_t is an LCS in which every unordered Cauchy series $\sum_{x\in S} x_s$ is absolutely Cauchy, and if L is the product space $\prod_{t\in T} L_t$, then in L every unordered convergent (or Cauchy) series is absolutely convergent (or Cauchy).

(7) (a) Dvoretsky has improved on the near-orthogonality which is given by Theorem 2. Call a number $\varepsilon > 0$ the *asphericity* of a symmetric convex body K in a Euclidean n-space l_n^2 with unit ball U if there are numbers $0 < \alpha \leq \beta \leq (1+\varepsilon)\alpha$ with $\alpha U \subseteqq K \subseteqq \beta U$.

Theorem 3. *If $\varepsilon > 0$ and k in ω are given, then there is an $n = n(k, \varepsilon)$ in ω so large that for every K in E_n there is a k-dimensional linear subspace L of l_n^2 for which the asphericity of $K \cap L$ is $< \varepsilon$.*

(b) Rephrased in terms of normed spaces this says: *For each infinite-dimensional normed space N and each n in ω and each $\varepsilon > 0$ there is a one-to-one linear mapping T of l_n^2, the n-dimensional Euclidean space, onto a subspace $N_{n,\varepsilon}$ of N such that $\|T\| \, \|T^{-1}\| < 1 + \varepsilon$.* In terms of Defn. VII, 4, 2, *every infinite-dimensional normed linear space mimics Hilbert space.* The current proofs are integral-geometric arguments which lead to very large estimates of $n(k, \varepsilon)$; the shortest one is in V. D. Mil'man [9]. See also Figiel.

§ 2. Tensor Products of Locally Convex Spaces

If L and M are linear spaces and L', M' are total linear subspaces of $L^{\#}, M^{\#}$ respectively, let $L \square M$ be the linear space whose basis elements are the elements of $L \times M$; that is, the ordered pairs $x \square y$ with x in L and y in M.

To each element $x \square y$ in $L \square M$ can be associated three functions:

(i) A bilinear functional $b = h_1(x, y)$ defined on $L' \times M'$ by $b(f, g) = f(x) g(y)$ for all f in L' and g in M';

(ii) A linear operator $u=h_2(x,y)$ from L' into M defined by $u(f)=f(x)y$ for all f in L'; and

(iii) A linear operator $v=h_3(x,y)$ from M' into L defined by $v(g)=g(y)x$ for all g in M'.

These mapping of each basis vector $x\square y$ in $L\square M$ determine three homomorphisms of the vector space $L\square M$ into three function spaces, $\mathfrak{b}(L',M')$, the space of all bilinear functionals on $L'\times M'$, $\mathfrak{L}(L',M)$, and $\mathfrak{L}(M',L)$. The kernel L_0 can be shown to be the same for these three homomorphisms. Define $L\bigcirc M$, the *tensor product* of L and M, to be $(L\square M)/L_0$, and let $x\bigcirc y$ be the image of $x\square y$ in $L\bigcirc M$. Then $L\bigcirc M$ is isomorphic to a space of bilinear functionals and to two spaces of linear operators.

(1) (a) Dually, $L'\bigcirc M'$ has three similar representations in $\mathfrak{b}(L,M)$, $\mathfrak{L}(L,M')$, and $\mathfrak{L}(M,L')$. (b) $L\bigcirc M$ and $L'\bigcirc M'$ are in duality (I, 2, (7)) under the bilinear form $\langle,\rangle$ defined at the basis elements by

$$\langle x\bigcirc y, f\bigcirc g\rangle = f(x)g(y).$$

(c) A set in $L\square M$ whose linear hull is L_0 is found from the elements $0\square y$, $x\square 0$, $(x_1+x_2)\square y-x_1\square y-x_2\square y$, $x\square(y_1+y_2)-x\square y_1-x\square y_2$, $(ax)\square y-a(x\square y)$, $(ax)\square y-x\square(ay)$, (where a is any real number).

Now suppose that L and M are LCS's; then there are many ways to induce a topology in $L\bigcirc M$ related to the given topologies in the original spaces and to the tensor product structure. Let us describe the normed case first, using Banach spaces A and B and their conjugate spaces A^* and B^*.

Definition 1. Let α and β be norms in A and B respectively; a *crossnorm of* α *and* β in $A\bigcirc B$ is a norm Θ in $A\bigcirc B$ such that $\Theta(x\bigcirc y)=\alpha(x)\beta(y)$ for all x in A, y in B. Define $\gamma=\gamma(\alpha,\beta)$, the *greatest crossnorm of* α *and* β, by

$$\gamma(u)=\inf\left\{\sum_{i\leq n}\alpha(x_i)\beta(y_i)\colon \sum_{i\leq n}x_i\bigcirc y_i=u\right\}.$$

(2) (a) γ is a norm in $A\bigcirc B$. (b) γ is a crossnorm of α and β. (c) If Θ is another crossnorm of α and β in $A\bigcirc B$, then $\Theta(u)\leq\gamma(u)$ for all u in $A\bigcirc B$. (d) γ is the Minkowski functional of the symmetric convex hull Γ of $U\bigcirc V=\{x\bigcirc y; x\in U$, the unit ball of α, and $y\in V$, the unit ball of $\beta\}$. Hence the natural mapping of $A\times B$, normed by $\|(x,y)\|=\sup\{\alpha(x),\beta(y)\}$, into $A\bigcirc B$ is of norm 1.

This γ is the greatest crossnorm attached to the norms α and β; there is no "least crossnorm" but there is a least crossnorm well fitted to this tensor product pattern.

Definition 2. If α and β are norms in the Banach spaces A and B, then $\lambda = \lambda(\alpha, \beta)$ is the norm imposed on $A \bigcirc B$ by any one of the three homomorphisms h_i; that is, if U and V are the unit balls of α and β, and if $u = \sum_{i \leqq n} x_i \bigcirc y_i$, then

$$\begin{aligned}\lambda(u) &= \sup\left\{\sum_{i \leqq n} f(x_i) g(y_i) : f \in U^\pi, g \in V^\pi\right\}\\ &= \sup\left\{\beta\left(\sum_{i \leqq n} f(x_i) y_i\right) : f \in U^\pi\right\}\\ &= \sup\left\{\alpha\left(\sum_{i \leqq n} g(y_i) x_i\right) : g \in V^\pi\right\}.\end{aligned}$$

Definition 3. Under the duality (1b) between $A \bigcirc B$ and $A^* \bigcirc B^*$, define the *dual norm Θ' of a crossnorm Θ* by $\Theta'(\varphi) = \sup\{\langle u, \varphi\rangle : \Theta(u) \leqq 1\}$ for each φ in $A^* \bigcirc B^*$.

(3) (a) λ is a crossnorm of α and β. (b) If α^* and β^* are the norms in A^* and B^*, and if $\lambda_0 = \lambda(\alpha^*, \beta^*)$, then if φ is an element of $A^* \bigcirc B^*$, its norm $\lambda_0(\varphi)$ can be computed either by looking at the associated bilinear functional on $A \times B$ or at one on $A^{**} \times B^{**}$; the two computations give the same value. (c) λ' is a crossnorm of α^* and β^* in $A^* \bigcirc B^*$ (d) If Θ is a crossnorm of α and β in $A \bigcirc B$, then Θ' is a norm in $A^* \bigcirc B^*$ if $\Theta'(f \bigcirc g) < \infty$ for every f in A^* and g in B^*. (e) Θ' is a crossnorm of α^* and β^* in $A^* \bigcirc B^*$ if and only if $\Theta \geqq \lambda(\alpha, \beta)$; then $\gamma(\alpha^*, \beta^*) \geqq \Theta' \geqq \lambda(\alpha^*, \beta^*)$. (f) $[\gamma(\alpha, \beta)]' = \lambda(\alpha^*, \beta^*)$.

This describes the critical property of λ [Schatten, page 32]; *λ is the least crossnorm of α and β whose dual norm is a crossnorm of α^* and β^*.*

(4) If L and M are LCS's, the topology in each space is determined by the family of continuous prenorms in the space. For such prenorms α and β, the Definitions 1 and 2 apply to give prenorms, rather than norms, in $L \bigcirc M$. These two families of prenorms define two locally convex topologies in $L \bigcirc M$, called the *projective* and the *weak tensor topologies* in $L \bigcirc M$. The completion of $L \bigcirc M$ in the projective (weak) tensor structure is denoted by $L \otimes M$ ($L \odot M$). (a) If A and B are Banach spaces, $A \odot B$ is isometric to the closure (in the topology of uniform convergence on $U^\pi \times V^\pi$) of the set of continuous bilinear functionals of finite rank. (b) Similarly, $L \odot M$ is the closure, in the topology of uniform convergence on the products of polars of neighborhoods (or on products of equicontinuous sets of functionals), of the set of those bilinear functionals of finite rank on $L^* \times M^*$ which are w^*-continuous in each variable.

(5) To show how this is connected with the theorem of Dvoretsky and Rogers, consider an index set S, its set Σ of finite subsets, a complete LCS L, and the spaces $l^1(S) \otimes L$ and $l^1(S) \odot L$. (a) If $\sigma \in \Sigma$ and if

$u=\sum_{i\leqq n} x_i \bigcirc y_i$, is an element of $l^1(\sigma)\bigcirc L$, then write $x_i=\sum_{s\in\sigma} x_{is}\delta_s$ and define $z_s=\sum_{i\leqq n} x_{is} y_i$; then u has a representation of the form $\sum_{s\in\sigma}\delta_s \bigcirc z_s$. (b) If α is the usual norm in $l^1(\sigma)$ and β is any continuous prenorm in L, let $\lambda_\beta=\lambda(\alpha,\beta)$ and $\gamma_\beta=\gamma(\alpha,\beta)$. Then $l^1(\sigma)\bigcirc L$ is isomorphic to $l^1(\sigma,L)$, the substitution space of Def. II, 2, 2; the isomorphism carries $\gamma_\beta(u)$ to $\sum_{s\in\sigma}\beta(z_s)$ and $\lambda_\beta(u)$ to $\sup\left\{\beta\left(\sum_{s\in\sigma} t_s z_s\right): \text{all } |t_s|\leqq 1\right\}$, where the z_s are computed from u as in (a). (c) If σ is part of S, think of $l^1(\sigma)$ as imbedded in $l^1(S)$ by setting the extra coordinates equal to zero. Then the preceding parts show that each element $u=\sum_{i\leqq n} x_i \circ y_i$ has a unique representation in terms of elements $z_s=\sum_{i\leqq n} x_{is} y_i$ where the series $\sum_{s\in S} z_s$ is unconditionally, and hence absolutely, convergent in the finite-dimensional subspace spanned by the elements y_i, $i\leqq n$. The same isometries still hold if sums over σ are replaced by sums over S; that is, $\gamma_\beta(u)=\sum_{s\in S}\beta(z_s)$ and $\lambda_\beta(u)=\sup\left\{\beta\left(\sum_{s\in S} t_s z_s\right): \text{all } |t_s|\leqq 1\right\}$. (d) Completing these spaces shows that $l^1(S)\otimes L$ is isomorphic to the set of all absolutely convergent series in L, u corresponds to $(z_s, s\in S)$ and $\gamma_\beta(u)=\sum_{s\in S}\beta(z_s)$. Similarly, $l^1(S)\odot L$ is isomorphic to the space of all unconditionally convergent series in L, u corresponds to $(z_s, s\in S)$ and $\lambda_\beta(u)=\sup\left\{\beta\left(\sum_{s\in S} t_s z_s\right): \text{all } |t_s|\leqq 1\right\}$.

The Dvoretsky-Rogers theorem of the preceding section combines with this result to show that when B is a Banach space $l^1(\omega)\otimes B$ fills up all of $l^1(\omega)\odot B$ if and only if B is finite-dimensional. Grothendieck [3] gives an alternative proof of this result and of related results for more general linear topological spaces. He shows that if M is a complete metrizable LCS such that $l^1(\omega)\otimes M=l^1(\omega)\odot M$, then M is nuclear; that is, $E\otimes M=E\odot M$ for all LCS's E. This is not true for all LCS's M but is also proved if M is what Grothendieck calls a space of type (DF), a family including Banach spaces and many others. R^ω is an example of a complete metrizable LCS in which unconditional and absolute convergence are equivalent while the space is not finite-dimensional.

(6) For the tensor product of a Hilbert space H with its adjoint H^*, $H\odot H^*$ can be interpreted, as usual, as the closure in $\mathfrak{L}(H)$ of the space of linear operators of finite rank; in this space the set $H\odot H^*$ is also the space (cc) of completely continuous operators belonging to $\mathfrak{L}(H)$. Schatten shows that $(H\odot H^*)^*$ is isometric to $H\otimes H^*$ and that $(H\odot H^*)^{**}$ is isometric to $\mathfrak{L}(H)$.

(7) Grothendieck [3] shows that the conjugate space of $L\otimes M$ is isomorphic to the space $\mathfrak{B}(L,M)$ of all bilinear functionals on $L\times M$

which are continuous in both variables simultaneously. The topology of $L \otimes M$ is then carried to the topology of uniform convergence on equicontinuous sets in $\mathfrak{B}(L, M)$. In general, a still stronger topology on $L \bigcirc M$ can be determined using as dual the space of all bilinear functionals on $L \times M$ which are continuous in the variables separately, using for neighborhoods of 0 in $L \bigcirc M$ the polar sets of separately equicontinuous sets.

(8) Ruston [1, 2] shows that the tensor product space is of great use in the study of the Fredholm theory of completely continuous operators. A part of the paper is given to extending the results of Schatten to products of n Banach spaces, in order to have the machinery to generalize Fredholm minors. Grothendieck also considers the application of tensor products to Fredholm theory in [3] and [4].

Schatten made the first long study of tensor products of normed spaces. Grothendieck [3] studied tensor products of locally convex spaces and invented nuclear spaces. The book of Pietsch is a more recent treatment of nuclear spaces. Schaefer [1] and Peressini deal with tensor products for spaces in which order is also important.

§ 3. Schauder Bases in Separable Spaces

In some (perhaps in all) separable, infinite-dimensional Banach spaces there exist sequences with special properties. These were first discussed by Schauder [1].

Definition 1. A *(Schauder) basis* (b_i) in a Banach space B is a sequence of elements of B such that for each x in B there is a unique sequence of real numbers (a_i), depending on x, such that $\lim_{n \in \omega} \left\| \sum_{i \leqq n} a_i b_i - x \right\| = 0$; the series $\sum_{i \in \omega} a_i b_i$ is called *the expansion of* x in the basis (b_i), and the coefficient $a_i = \beta_i(x)$ is the ith *coordinate of* x in the basis (b_i).

During this chapter we shall abbreviate by using only the word basis for "Schauder basis"; there will be no occasion here to use the vector or Hamel basis of Chapter I.

Definition 2. If (b_i) is a basis for B, let $U_m x = \sum_{i \leqq m} \beta_i(x) b_i$ and let $V_m x = x - U_m x$. (b_i) is called a *monotone basis* for B if for each x in B, $\|U_m x\|$ is a non-decreasing function of m.

(1) (a) If x and $y \in B$, then $\beta_i(x) + \beta_i(y)$ is a suitable coordinate sequence for $x + y$; by uniqueness of the expansion each β_i is an additive functional on B. Similarly it can be shown that each β_i is homogeneous.

(b) Hence U_m and V_m are linear. (c) A basis (b_i) is monotone if and only if $\|U_m\| \leqq 1$ for each $m \in \omega$.

Theorem 1. *Let (b_i) be a basis for B, define a new norm for the space by $\|x\|' = \sup\{\|U_m x\| : m \in \omega\}$, and let B' be the same vector space with the new norm. Then*

(i) *B' is a Banach space isomorphic to B.*

(ii) *(b_i) is a monotone basis for B'.*

(iii) *Each β_i is the unique element of B^* such that $\beta_i(b_j) = \delta_{ij}$* (Kronecker's delta) *for all i,j in ω; also for each i, $\|\beta_i\|' \, \|b_i\|' \leqq 2$.*

(iv) *$U_m U_n = U_p$, where $p = \min(m,n)$, so that each U_m is idempotent; U_m and V_m are continuous linear operators in B (or in B').*

(v) *$\|U_m\|' \leqq 1$, so $\|V_m\|' \leqq 2$, and there is a K with $\|U_m\| \leqq K$ and $\|V_m\| \leqq K+1$.*

(vi) *If $\beta \in B^*$, then $U_m^* \beta$ (which $= \sum_{i \leqq m} \beta(b_i)\beta_i$) is w*-convergent to β; that is, (β_i) is a w*-basis in B^*.*

Proof. Each function $p_m(x) = \|U_m x\|$ is sublinear; for each x, $\lim_{m \in \omega} U_m x = x$ so $\|x\|' = \sup_m p_m(x) < \infty$ and it at least as great as $\|x\|$ for each x. Hence $\|\dots\|'$ has the properties of a norm and B' is a normed linear space.

To show B' is complete take a Cauchy sequence (x_n) in B'; then $x_n = \sum_{i \in \omega} \beta_i(x_n) b_i$. For each k and each $j > k$ let $U_{kj} = U_j - U_k$; then

$$\|U_{kj}(x_m - x_n)\| \leqq \|U_j(x_m - x_n)\| + \|U_k(x_m - x_n)\| \leqq 2\|x_m - x_n\|',$$

in particular,

$$|\beta_j(x_m - x_n)| \, \|b_j\| = \|\beta_j(x_m - x_n) b_j\| = \|U_{j-1,j}(x_m - x_n)\| \leqq 2\,\|x_m - x_n\|'$$

for each j. Hence the sequence $(\beta_j(x_m) b_j, m \in \omega)$ converges in B, say to the point $a_j b_j$.

Given $\varepsilon > 0$ there exists m_ε such that $n > m > m_\varepsilon$ imply $\|x_m - x_n\|' < \varepsilon$. Taking the limit on n shows that $\left\| U_{kj} x_m - \sum_{k < i \leqq j} a_i b_i \right\| \leqq 2\varepsilon$ for all $k < j$ and $m > m_\varepsilon$. For fixed $m > m_\varepsilon$, $\lim_{k \in \omega} \|U_k x_m - x_m\| = 0$, so there exists k_ε so large that $\|U_{kj} x_m\| < \varepsilon$ when $j > k > k_\varepsilon$; therefore $\left\| \sum_{k < i \leqq j} a_i b_i \right\| < 3\,\varepsilon$ when $j > k > k_\varepsilon$. The completeness of B implies that $\sum_{i \in \omega} a_i b_i$ converges to some element x of B; then by uniqueness of the expansion of x, $a_i = \beta_i(x)$ and $U_{kj} x = \sum_{k < i \leqq j} a_i b_i$. But then $\|x_m - x\|' = \sup_k \|U_k x_m - U_k x\| \leqq \varepsilon$ when $m > m_\varepsilon$; hence B' is also complete.

By II, 3, (8), the identity mapping is an isomorphism of B with B'; this proves (i). Then there is a $k > 0$ such that $k\|x\| \geqq \|x\|' \geqq \|x\|$. This means that to prove (v) it suffices to prove that $\|U_m\|' \leqq 1$; for this the easily verified rules of calculation stated in (iv) are required. Then $\|U_m\|' = \sup\{\|U_m x\|' : \|x\|' \leqq 1\} = \sup\{\|U_k U_m x\| : \|x\|' \leqq 1$ and $k \in \omega\} = \sup \{\|U_k x\| : \|x\|' \leqq 1$ and $k \leqq m\}$. But $\|x\|' \leqq 1$ means $\|U_k x\| \leqq 1$ for all k in ω; hence $\|U_m\|' \leqq 1$. (ii) follows from this and (1c).

For each x, $|\beta_n(x)| \, \|b_n\|' = \|\beta_n(x) b_n\|' = \|U_n x - U_{n-1} x\|' \leqq 2\|x\|'$ so $\|\beta_n\|' \, \|b_n\|' \leqq 2$. This proves that each β_i is in $B^* = B'^*$. Uniqueness of the expansion of b_j proves that $\beta_i(b_j) = \delta_{ij}$, so (iii) holds. (vi) now follows from direct calculation: $\beta(b) = \lim_{m \in \omega} \beta(U_m b) = \lim_{m \in \omega} (U_m^* \beta)(b)$ for every b in B.

Schauder [1] assumed that $\|b_i\| = 1$ and that each $\beta_i \in B^*$; this proof that the β_i are all continuous is due to Banach [p. 111]. Clearly a change from b_i to $b_i' = b_i / \|b_i\|$ is allowable if at the same time β_i is replaced by $\beta_i' = \|b_i\| \, \beta_i$, so the basis elements may be normalized without losing any of the properties of Theorem 1.

(2) In $c_0(\omega)$ and in $l^p(\omega)$, $p \geqq 1$, the sequence $(b_i, i \in \omega)$, for which $b_i = (\delta_{ij}, j \in \omega)$ for each i, is a basis.

(3) In $C[0,1]$ Schauder constructed a basis of polygonal functions: $b_0(t) = t$; $b_1(t) = 1 - t$; $b_2(t) = 0$ if $t \leqq 0$, $= 2b_0(t)$ if $0 \leqq t \leqq 1/2$, $= 2b_1(t)$ if $1/2 \leqq t \leqq 1$, $= 0$ if $t \geqq 1$; for $i = 1$ or 2, $b_{2+i}(t) = b_2(2t - i + 1)$; for $i = 1, 2, 3, 4$, $b_{4+i}(t) = b_2(4t - i + 1)$; ...; for $i = 1, \ldots, 2^n$, $b_{2^n + i}(t) = b_2(2^n t - i + 1)$; ... Then the β's are described for each x in $C[0,1]$ by $\beta_0(x) = x(1)$; $\beta_1(x) = x(0)$; $\beta_2(x) = x(1/2) - (x(0) + x(1))/2$; and so on by induction.

(4) Call sequences (b_i) in B and (β_i) in B^* *biorthogonal* if $\beta_i(b_j) = \delta_{ij}$. Then (b_i) is a basis for B if and only if (i) there is a biorthogonal sequence (β_i) in B^*, and (ii) $(b_i, i \in \omega)$ spans B, and (iii) there is a K such that $\left\| \sum_{i \leqq m} \beta_i(x) b_i \right\| \leqq K$ if $\|x\| \leqq 1$ and $m \in \omega$. [Banach, p. 107, Th. 3.]

Theorem 2. *Let* (b_i) *be a sequence of elements of* B *such that for each* x *there is a unique sequence* $a_i = \beta_i(x)$ *of real numbers such that* (i) $\sum_{i \in \omega} a_i b_i$ *converges weakly to* x *and* (ii) *each* $\beta_i \in B^*$ *or* (ii') B *is w-sequentially complete. Then* (b_i) *is a basis for* B *and* (β_i) *is a basis for the closed linear subspace* Γ *of* B^* *spanned by the* β_i. (That is to say, a weak basis in a Banach space is a norm basis.)

Proof. Defining U_m as before, we have $\lim_{m \in \omega} \beta(U_m x) = \beta(x)$ for all β in B^* and x in B; (ii) implies continuity of each U_m; the category theorems, [II, § 3, (1d)] assert that $\|U_m\|$ is uniformly bounded. Then (4) could be applied, but a direct proof follows from Lemma II, 3, 1 which asserts that the set E of x for which the $U_m x$ converge in the norm topology is closed and linear. But $\|U_m b_j - b_j\| = 0$ when $m \geqq j$, so E contains all b_j; since E is weakly closed, it contains all of B. The same argument applies to the $U_m^* \beta_i$.

If instead of (ii) the condition (ii′) holds, follow the proof of Theorem 1 down to the completeness of B' to get continuity of each β_i, that is, (ii). (w-convergence and w-completeness of B are used in the third paragraph, where norm completeness of B sufficed under the stronger hypotheses of Theorem 1.)

Singer, p. 209, reports that Bessaga and Pelczynski have proved Banach's statement that (i) suffices for this conclusion.

Definition 3. A basis (b_i) of B is called *boundedly complete* in B if for each sequence (a_i) of real numbers such that the sequence $\left(\left\|\sum_{i\leqq n} a_i b_i\right\|, n\in\omega\right)$ is bounded in B there is an x in B such that $a_i=\beta_i(x)$ for all i, so $x=\sum_{i\in\omega} a_i b_i$. A basis (b_i) is called a *shrinking basis* for B if for each β in B^*, $\lim_{n\in\omega} p_n(\beta)=0$ where $p_m(\beta)=$norm of β restricted to the range of V_m; that is, $p_m(\beta)=\sup\{\beta(x)\colon x=V_m x \text{ and } \|x\|\leqq 1\}$.

Next we investigate when (β_i) is a basis, not just a w^*-basis, for all of B^*.

Lemma 1. *If (b_i) is a basis for B, then the following conditions are equivalent:*

(i) *(β_i) is a basis for B^*.*

(ii) *$(V_m^*, m\in\omega)$ tends to zero in the strong operator topology.*

(iii) *(b_i) is a shrinking basis for B.*

(iv) *(β_i) spans B^*.*

Proof. From Theorem 1, (vi), $(\beta(b_i), i\in\omega)$ is the only possible coefficient sequence for β; hence (β_i) is a basis for B^* if and only if $\|V_m^*\beta\|\to 0$ for all β, so (i) and (ii) are equivalent. (iv) and (i) are equivalent by the last part of Theorem 2.

$$\begin{aligned}\text{Clearly } p_m(\beta)&=\sup\{\beta(x)\colon x=V_m x \text{ and } \|x\|\leqq 1\}\\ &\leqq \sup\{\beta(V_m x)\colon \|x\|\leqq 1\}\\ &=\sup\{V_m^*\beta(x)\colon \|x\|\leqq 1\}=\|V_m^*\beta\|.\end{aligned}$$

But $\{V_m x\colon \|x\|\leqq 1\}\subseteqq\{x\colon x=V_m x \text{ and } \|x\|\leqq K\}$ where K is a common bound for $\|V_m\|$, so $\|V_m^*\beta\|\leqq K p_m(\beta)$. Hence (ii) and (iii) are equivalent.

Lemma 2. *Let (b_i) be a monotone basis for B, let (β_i) be the corresponding biorthogonal sequence in B^*, and let Γ be the closed linear space spanned by the β_i. Then* (i), (ii), *and* (iii) *below are equivalent and imply* (iv):

(i) *(b_i) is a boundedly complete basis.*

(ii) *For each F in B^{**} the series $\sum_{i\in\omega} F(\beta_i) b_i$ converges to a point y_F of B with $\|y_F\|\leqq\|F\|$.*

(iii) $QB \cap \Gamma^{\perp} = \{0\}$ *and the projection* T *of* B^{**} *onto* QB *along* $\Gamma^{\perp}$ *is of norm* 1.

(iv) B *is isometric to a conjugate space* $(=\Gamma^*)$.[1]

Proof. If (i) holds, take F in B^{**} with $\|F\| \leqq 1$. Then by Theorem II, 5, 4 there exists a sequence $(y_k, k \in \omega)$ such that $\|y_k\| \leqq 1$ for each k and $\lim_{k\in\omega} \beta_i(y_k) = F(\beta_i)$ for all i in ω; hence $\lim_{k\in\omega} \gamma(y_k) = F(\gamma)$ for every γ in Γ. Then for each n, $\sum_{i \leqq n} F(\beta_i) b_i = \lim_{k\in\omega} \sum_{i\leqq n} \beta_i(y_k) b_i = \lim_{k\in\omega} U_n y_k$. But $\|U_n\| \leqq 1$, so $\left\| \sum_{i\leqq n} F(\beta_i) b_i \right\| = \lim_{k\in\omega} \|U_n y_k\| \leqq \lim_{k\in\omega}\sup \|U_n\| \, \|y_k\| \leqq 1$. By (i) there exists y_F in B satisfying (ii).

If (ii) holds, let $TF = Q y_F$; then T is linear and $\|TF\| = \|y_F\| \leqq \|F\|$ if $F \in B^{**}$. Also $F - TF$ vanishes on each β_i; therefore $F - TF \in \Gamma^{\perp}$. If $G \in \Gamma^{\perp}$, then $TG = 0$, so $T^2 = T$; also, if $TF = 0$, all $F(\beta_i) = 0$ or $F \in \Gamma^{\perp}$. Hence T is a projection of norm 1 of B^{**} on QB along $\Gamma^{\perp}$. If (iii) holds, let (a_i) be a sequence of real numbers such that if $x_n = \sum_{i\leqq n} a_i b_i$, then $(x_n, n \in \omega)$ is bounded, say $\|x_n\| \leqq 1$ for all n in ω. Then $\lim_{n\in\omega} \beta_i(x_n) = a_i$ for all i so [Th. II, 3,2] for every γ in Γ, $\lim_{n\in\omega} \gamma(x_n) = \varphi(\gamma)$ exists; φ is in Γ^*, and $\|\varphi\| \leqq 1$. Take F in B^{**} so that $\|F\| = \|\varphi\|$ and F is an extension of φ; let $y = Q^{-1} TF$. Then $F - TF \in \Gamma^{\perp}$, so $\gamma(y) = TF(\gamma) = \varphi(\gamma) = \lim_{n\in\omega} \gamma(x_n)$; in particular, $\beta_i(y) = \lim_{n\in\omega} \beta_i(x_n) = a_i$. Hence the expansion of y in the basis (b_i) is $\sum_{i\in\omega} a_i b_i$ and (iii) implies (i).

B is isometric to QB which by (iii) is isometric to $B^{**}/\Gamma^{\perp}$. But by Lemma II, 1,2, $B^{**}/\Gamma^{\perp}$ is isometric to Γ^*, so (iii) implies (iv).

Turn now to the connections between bases and reflexivity; much of the rest of this work comes directly from R. C. James [1, 2].

Theorem 3. *If* (b_i) *is a basis for* B, *then* B *is reflexive if and only if the basis is both shrinking and boundedly complete.*

Proof. Suppose that B is reflexive. Then the w- and w^*-topologies agree in B^{**}; by Theorem 1, (vi) and Theorem 2, (β_i) is a basis for B^* and by Lemma 1, (b_i) is shrinking. Since $\Gamma^{\perp} = \{0\}$, Theorem 1, (ii), and Lemma 2 imply that (b_i) is boundedly complete. If on the other hand, (b_i) is shrinking and boundedly complete, by Lemma 1, $\Gamma = B^*$, so $\Gamma^{\perp} = \{0\}$. By Lemma 2, $QB = B^{**}$.

---

[1] Alaoglu observed that (i) implies (iv).

Corollary 1. *Let (b_i) be a shrinking basis for B and let (β_i) be the corresponding basis for B^*; then* (i) *(β_i) is a monotone basis if and only if (b_i) is monotone basis, and* (ii) *(β_i) is a boundedly complete basis.*

Proof. (i) follows from Lemma 1 and (1). (ii) can be proved as "(iii) implies (i)" of Lemma 2 was proved, or as follows: Let $B' = B^*$; then (β_i) is a basis and $(Q b_i)$ is the corresponding biorthogonal system in $B'^* = B^{**}$, so $\Gamma' = QB$. $Q' Q^*$ is the projection of $B^{***} = B'^{**}$ onto $Q' B'$ along $\Gamma'^{\perp}$ and is of norm 1. By "(iii) implies (i)" of Lemma 2, (β_i) is a boundedly complete basis for $B^* = B'$.

The next theorem describes B^{**} when (b_i) is a shrinking basis.

Theorem 4. *Let (b_i) be a monotone shrinking basis for B and let (β_i) be the corresponding boundedly complete basis for B^*; then:*

(i) *Each sequence of numbers (d_i) such that $\sup_{n \in \omega} \left\| \sum_{i \leqq n} d_i b_i \right\| < \infty$ determines an element F of B^{**} by means of the relation $F(\beta) = \sum_{i \in \omega} d_i c_i$ whenever $\beta = \sum_{i \in \omega} c_i \beta_i$.*

(ii) *If $F \in B^{**}$ and $d_i = F(\beta_i)$ for every i in ω, then $\|F\| = \lim_{n \in \omega} \left\| \sum_{i \leqq n} d_i b_i \right\|$ and $F(\beta) = \sum_{i \in \omega} d_i c_i$ whenever $\beta = \sum_{i \in \omega} c_i \beta_i$.*

Proof. Most of this conclusion is in the assertion that $(Q b_i, i \in \omega)$ is a w^*-basis for B^{**}. This follows from Theorem 1, (vi), lifted one space, for Lemma 1 assures us that (β_i) is a basis for B^* and that its sequence of coefficient functionals is $(Q b_i)$.

We give next an example of R. C. James [2] which displays a non-reflexive space isometric to its second conjugate space although *not* under the natural isometry; the deficiency of QB in B^{**} is precisely 1.

Example. If Σ is the set of finite subsets σ of the integers, define functions T from R^{ω} into $l^2(\omega)$ as follows: If $x = (x(i), i \in \omega)$, and σ consists of the integers $i_i < i_2 < \cdots < i_n$, then

$$\begin{aligned} T_\sigma x(j) &= x(i_{j+1}) - x(i_j) \text{ if } 1 \leqq j < n, \\ T_\sigma x(n) &= x(i_1) - x(i_n), \\ T_\sigma x(j) &= 0 \text{ if } j > n. \end{aligned}$$

Set $p_\sigma(x) = \| T_\sigma x \|_{l^2}$ and let B be the normed space of all those sequences x in R^{ω} such that

(i) $\|x\| = \sup\{p_\sigma(x) : \sigma \in \Sigma\} < \infty$, and

(ii) $\lim_{i \in \omega} x(i) = 0$.

(5) (a) (i) implies that $\lim_{i \in \omega} x(i)$ exists. (b) Setting $i_1 = k$ and letting $i_2 \to \infty$, (i) and (ii) imply that $\|x\|^2 \geqq 2|x(k)|^2$ for each x in B and k in ω.

(c) Under this norm B is a Banach space. (d) If E_i is the subspace of B containing just those elements x for which $x(k)=0$ if k is congruent to i mod 2, then each E_i is isomorphic to $l^2(\omega)$, and is closed in B; $E_1 \cap E_2 = \{0\}$, but, James [2] to the contrary, $E_1 + E_2$ is not all of B.

James [2] shows that the unit vectors in B form a monotone, shrinking, but *not* boundedly complete basis in B. From Theorem 4 he shows that B^{**} is isometric to the space of sequences satisfying condition (i), and that $Q(B)$ is embedded in it in the original way as the space of sequences satisfying both conditions (i) and (ii); hence the deficiency of $Q(B)$ in B^{**} is precisely one.

The isometry between B^{**} and B is set up as follows: If X in B^{**} corresponds to a sequence $(d_i, i \in \omega)$ satisfying (i), let d be the limit of (d_i) and let $TX = (x_i)$ where $x_1 = -d$, and $x_{i+1} = d_i - d$ for $i = 1, 2, \ldots$

The literature on bases has grown enormously since 1958. Fortunately there exists a comprehensive account of bases and related systems by Singer; for those interested in applications of bases there is the excellent survey by V. D. Mil'man [10] in Russian or in English translation.

Various results are known about *basic sequences*, that is, sequences which are bases for the subspaces they span. For example, (Day [15]) from a point-to-set version of Borsuk's antipodal-point theorem one can prove that for each finite-dimensional subspace E of N and each subspace F of larger dimension than E that there is x in F such that $\|x+E\| = \|x\|$. (See Kreĭn-Krasnoselski-Mil'man whose result is more accessible in Gohberg-Kreĭn, Th. 1.1 and Cor.) From this it is possible to get Day [15].

Theorem 5. *If N is an infinite dimensional normed space and if* (c_m) *is a sequence of positive numbers, then there exists biorthogonal sequences* (b_i) *in N and* (β_i) *in N^* such that* (i) $\|b\| = 1 = \|\beta_i\|$ *for all i in* ω, (ii) (b_i) *is a basis for the closed linear manifold L which it spans in B, and* (iii) *if* $U_m(x) = \sum_{i \leqq m} \beta_i(x) b_i$, *then* U_m *is a projection in L of norm* $\leq 1 + c_m$. [Note that, from the proof in Day [15], if N is so large that there is no countable total subset in N^*, then the c_m can all be taken to be 0.]

V. D. Mil'man [10] calls a basis in which $\|b_i\| = 1 = \|\beta_i\|$ for all i in ω an Auerbach *basis* and a basis in which $\|U_m\| \to 1$ an *asymptotically monotone* basis. He gives many conditions on a sequence (x_i) sufficient to make it possible to select a basic sequence from (x_i), and illustrates how many topological properties of normed spaces follow from these ways to choose bases for subspaces.

The Approximation problem: Enflo's counterexample. We close this section with a report on the recent negative answer to Schauder's question:

Does every separable Banach space have a basis [Banach, p. 111]. This requires some description of work of Grothendieck [3, Chap. 1, § 5.] The example of Enflo is too complicated to give here but is to be found in Enflo [2].

In his analysis of tensor products Grothendieck isolated an important property which all the familiar Banach spaces have. The several variants are successively stronger restrictions on B; for separable spaces the bounded approximation property is formally weaker than the presence of a Schauder basis.

Definition 4. An LCS L, in particular, a normed space, is said to have the *approximation property* (AP) if there is a net $(T_n, n \in \Delta)$ of continuous linear operators of finite rank in L such that (T_n) converges to the identity operator uniformly on each totally bounded set K in L. A normed space N has the *bounded approximation property* (BAP) if there is such a net with a uniform bound on $\|T_n\|$; N has the *metric approximation property* (MAP) if there is such a net with $\|T_n\| \leq 1$ for all n in Δ.

(6) Grothendieck [3, page 167] lists a number of properties of a space B equivalent to some of these, mostly involving tensor products of B with other spaces, and he uses these to prove: (a) If B^* has (AP), so does B. (b) B^* has (AP) if and only if for each Banach space F, each compact linear map T of B into F is the limit in the norm topology of $L(B,F)$ of elements of $B^* \otimes F$; that is, of operators of finite rank from B into F. (c) On p. 180 Grothendieck shows simply that every $L^p(\mu)$, $1 \leq p \leq \infty$, has (MAP). Then by (a) every abstract L-space and abstract M-space (VI, § 4) has (MAP). (c) Clearly every Banach space with a basis has (BAP).

(7) Grothendieck [3, page 170] then gives a number of conjectures equivalent to the conjecture: *Every Banach space has the approximation property*. Some of these (p. 171) are quite special properties of particular spaces, such as his (f) for every compact measure space (X,μ) and every continuous function K on $X \times X$ such that $\int_X K(x,s)K(s,y)\,d\mu(s)=0$ it follows that $\int_X K(s,s)\,d\mu(S)=0$, or (f''). If $u \in l^1 \times c_0$ and $u^2=0$, then the trace of u is 0.

(8) (a) Grothendieck [3, p. 181] shows that if a reflexive space has (AP), then B and B^* have (MAP). (b) Rosenthal observed that if B^* is a separable conjugate space with (AP), then B and B^* have (BAP).

(9) Enflo [2] has constructed a subspace of c_0 which does not have (BAP). By (8b) the dual B^*, which is a factor space of l^1 and is therefore separable, does not have (AP). These spaces B and B^* are counter examples to show also that separable spaces need not have a basis.

(10) Pelczynski [3] has proved that a separable B has the bounded approximation property if and only if it is a complemented subspace of a space with a Schauder basis.

§ 4. Unconditional Bases[2]

Definition 1. A basis $(b_i, i \in \omega)$ in B is called an *unconditional (absolute) basis* for B if and only if for each x the series $\sum_{i\in\omega} \beta_i(x) b_i$ is unconditionally (absolutely) convergent to x.

The first theorem of this section is analogous to that of the preceding section. If Σ is the set of finite subsets v of ω, define $U_v\, x = \sum_{i\in v} \beta_i(x) b_i$ and $U_\emptyset\, x = 0$ for all x. Then define $V_v = \mathrm{i} - U_v$ and $W_v = \mathrm{i} - 2U_v = V_v - U_v$.

Theorem 1. *Let $(b_i, i \in \omega)$ be an unconditional basis for B; let $\|x\|' = \sup\{\|U_v\, x\| : v \in \Sigma\}$; let $\|x\|'' = \sup\{\|W_v\, x\| : v \in \Sigma\}$. Let B' and B'' be the spaces obtained by renorming B with $\|\ldots\|'$ and $\|\ldots\|''$. Then*

(i) *B' and B'' are Banach spaces isomorphic to B.*

(ii) *Every rearrangement of (b_i) is a monotone basis for B' and for B''.*

(iii) *Each β_i is the unique element of B^* for which $\beta_i(b_j) = \delta_{ij}$ for all j in ω, and $\|\beta_i\|'\, \|b_i\|' = 1$ for all i in ω.*

(iv) *If μ and $v \in \Sigma$, then $U_\mu U_v = U_{\mu \cap v}$, so for each μ, $U_\mu^2 = U_\mu$ and $W_\mu^2 = i$. Setting $\lambda =$ symmetric difference[3] of μ and v, $W_\mu W_v = W_\lambda$.* (This implies that every U_μ and every V_μ is a projection and that every W_μ is an involution[4] in B.)

(v) *$U_\mu x$ tends to x and $V_\mu x$ tends to zero for each x in B. Also $\|U_\mu\|' \leqq 1$ and $\|W_\mu\|'' \leqq 1$, so $\|U_\mu\|'' \leqq 1$ and $\|V_\mu\|'' \leqq 1$ for all μ in Σ. If μ and v are disjoint and non-empty, if $x = U_\mu x \neq 0$, and if $y = U_v\, y \neq 0$, then $\|x + t\,y\|''$ is an even, convex[5] function of t, so it is a non-decreasing function of $|t|$.*

(vi) *$(\beta_i, i \in \omega)$ is a w^*-unconditional basis for B^*; that is, for each β in B^* the series $\sum_{i\in\omega} \beta(b_i)\beta_i$ is w^*-unordered (even w^*-absolute) convergent to β.*

The difficult part of this proof follows the pattern of Theorem 1 of the preceding section. The rest follows by exploiting the involutions W_μ, $\mu \in \Sigma$.

---

[2] This terminology is that of James [1,2], not that of Karlin [1] or Gelbaum. By § 1, (1d) any of the definitions (*B*)—(*D*) can be used to define unconditional convergence. We shall generally use unordered convergence.

[3] The symmetric difference of two sets is the union minus the intersection.

[4] An involution is a linear operator whose square is the identity.

[5] If a Banach space has a basis with this property, then the expansion of each element is unordered Cauchy, so the basis is unconditional.

For unconditional bases, there are alternative conditions equivalent to bounded completeness or shrinking. The usual bases in $c_0(\omega)$ and in $l^p(\omega)$, $p \geqq 1$, are unconditional; that defined by Schauder [1] in $C[0,1]$ is conditional (see Corollary 1) and the basis in James' example at the end of the preceding section achieves its purpose because it is conditional.

Theorem 2. *If (b_i) is an unconditional basis for B, then the following conditions are equivalent:*

(a) *The basis is boundedly complete.*

(b) *The space is weakly sequentially complete.*

(c) *There is no subspace of B isomorphic to $c_0(\omega)$.*

Proof. Without loss of generality, we may assume that the basis and the norm in B have the relationships expressed in Theorem 1, (v), for $\|\ldots\|''$; this might be called unconditional monotony of the basis.

To prove (c)⇒(a) suppose that there is a sequence (a_i) such that $\left(\sum_{i \leqq n} a_i b_i, n \in \omega\right)$ is a bounded sequence in B which does not converge to any element of B. Then it can be assumed that $\left\|\sum_{i \in \mu} a_i b_i\right\| < 1$ for every μ in Σ. Since the series is not Cauchy in B, there exist sequences of integers (n_k) and (m_k) and a positive number d such that $n_k < m_k < n_{k+1}$ for every k in ω and, setting z_k = sum of all $a_i b_i$ with $n_k \leqq i < m_k$; $\|z_k\| \geqq d$ for all k.

It suffices to show that $(z_k, k \in \omega)$ is a basis for a subspace of B isomorphic to $c_0(\omega)$. Take $\mu \in \Sigma$ and real numbers t_k, $k \in \mu$; by (v) of Theorem 1, $\left\|\sum_{k \in \mu} t_k z_k\right\|$ is a non-decreasing function of each $|t_k|$, $k \in \mu$. Hence $\left\|\sum_{k \in \mu} t_k z_z\right\| \leqq (\sup\{|t_k| : k \in \mu\}) \left\|\sum_{k \in \mu} z_k\right\| \leqq \sup\{|t_k| : k \in \mu\}$ for if $v = \bigcup_{k \in \mu} \{i : n_k \leqq i < m_k\}$, then

$$\left\|\sum_{k \in \mu} z_k\right\| = \left\|\sum_{i \in v} a_i b_i\right\| < 1.$$

But also by Theorem 1, (v),

$$\left\|\sum_{k \in \mu} t_k z_k\right\| \geqq \sup\{\|t_k z_k\| : k \in \mu\} = \sup\{|t_k| \cdot \|z_k\| : k \in \mu\} \geqq d \sup\{|t_k| : k \in \mu\}.$$

Therefore, if T carries each finite linear combination x of the basis vectors in $c_0(\omega)$ to the same combination of the z_k, then the distortion of the norm is bounded; $\|x\| \geqq \|Tx\| \geqq d\|x\|$. It follows that this correspondence T can be extended by continuity to an isomorphism of $c_0(\omega)$ into the closed linear manifold in B spanned by (z_k).

To prove (a)⇒(b) assume (b_i) boundedly complete, and take a sequence $(x_n, n \in \omega)$ in B such than $\lim_{n \in \omega} \beta(x_n)$ exists for every β in B^*;

then the category theorems, Cor. II, 3,1, assert that there is a K with $\|x_n\| \leqq K$ for all n. Let $a_i = \lim_{n\in\omega} \beta_i(x_n)$; then for each μ in Σ,

$$\lim_{n\in\omega}\Big\| U_\mu x_n - \sum_{i\in\mu} a_i b_i \Big\| = \lim_{n\in\omega}\Big\| \sum_{i\in\mu} (\beta_i(x_n) - a_i) b_i \Big\|$$
$$\leqq \lim_{n\in\omega} \sum_{i\in\mu} |\beta_i(x_n) - a_i| \, \|b_i\| = 0.$$

Hence $\Big\| \sum_{i\in\mu} a_i b_i \Big\| \leqq \limsup_{n\in\omega} \|U_\mu x_n\| \leqq \limsup_{n\in\omega} \|x_n\| \leqq K$ for all μ. By bounded completeness there is an x in B such that $a_i = \beta_i(x)$.

Setting $y_n = x_n - x$, we need only show that y_n tends weakly to 0. If not, then there exists β of norm 1 in B^*, $\varepsilon > 0$, and a sequence (n_m) such that $\beta(y_{n_m}) > \varepsilon$ for all m. Let $z_m = y_{n_m}$ then also $\lim_{m\in\omega} \|U_\mu z_m\| = 0$ for each μ. Then take $\eta = \varepsilon/6$; there exist increasing sequences (n_k) and (m_k) such that

$$\|V_{m_1} z_{n_1}\| < \eta, \quad \|U_{m_1} z_n\| < \eta \quad \text{if } n \geqq n_2, \quad \|V_{m_2} z_{n_2}\| < \eta, \quad \text{and so on.}$$

Let

$$w_k = U_{m_k} V_{m_{k-1}} z_{n_k}; \quad \text{then} \quad \|w_k - z_{n_k}\| < 2\eta$$

and

$$\beta(w_k) = \beta(z_{n_k}) - \beta(z_{n_k} - w_k) > \varepsilon - 2\eta.$$

Then if $\tau = \sum_{k\in\rho} t_k \delta^k$, where $\rho \in \Sigma$, set $T\tau = \sum_{k\in\rho} t_k w_k$; then

$$\|T\tau\| = \Big\| \sum_{k\in\rho} t_k w_k \Big\| \leqq \|\tau\|_{l^1} \sup\{\|w_k\| : k \in \rho\} < 2(K + \eta)\|\tau\|_{l^1}.$$

But by Theorem 1, (v)

$$\Big\| \sum_{k\in\rho} t_k w_k \Big\| = \Big\| \sum_{k\in\rho} |t_k| w_k \Big\| \geqq \Big|\beta\Big(\sum_{k\in\rho} |t_k| w_k\Big)\Big| = \sum_{k\in\rho} |t_k| \beta(w_k)$$
$$\geqq (\varepsilon - 2\eta) \sum_{k\in\rho} |t_k| = (\varepsilon - 2\eta)\|\tau\|_{l^1}.$$

Now if $T'\tau = \sum_{k\in\rho} t_k z_{n_k}$, then $\|T'\tau - T\tau\| < 2\eta\|\tau\|_{l^1}$, so

$$\|\tau\|_{l^1}(2K + 4\eta) \geqq \|T'\tau\| \geqq (\varepsilon - 4\eta)\|\tau\|_{l^1}.$$

Hence T' determines an isomorphism of $l^1(\omega)$ into B such that $T'\delta^k = z_{n_k}$. If $B_1 = T'(l^1(\omega))$, then B_1 is w-ω-complete, because $l^1(\omega)$ is. Because $\lim_{k\in\omega} \beta(z_{n_k}) = \lim_{n\in\omega} \beta(y_n)$ exists for all β in B^*, (z_{n_k}) converges weakly to some element z of B_1; but $\beta_i(z) = \lim_n \beta_i(y_n) = 0$ for all i in ω, so $z = 0$, and 0 is the weak limit of the y_n after all. This contradiction shows that bounded completeness of (b_i) implies w-ω-completeness of B.

To prove that (b) implies (c) one need only observe that, by the Hahn-Banach theorem, every norm- or w-closed linear subspace of a w-ω-complete space is itself w-ω-complete. Since $c_0(\omega)$ is not w-ω-complete, for it is w^*-ω-dense in $m(\omega)$, it follows that no isomorphic image of $c_0(\omega)$ can appear in a w-ω-complete space.

Theorem 3. *Let (b_i) be an unconditional basis for B; then the basis is shrinking if and only if there is no subspace of B isomorphic to $l^1(\omega)$.*

Proof. If necessary the space can be renormed isomorphically to have the property of Theorem 1, (v). Suppose that (b_i) is not a shrinking basis. By Theorem 1, (v) and Theorem 3,1, there exist $\varepsilon>0$, $\beta\in B^*$, indices m_k increasing indefinitely, and points y_k in B such that $\|\beta\|=\|y_k\|=1$, $\beta(y_k)>\varepsilon$, $V_{m_k}y_k=y_k$, and $V_{m_{k+1}}y_k=0$. If $\mu\in\Sigma$, and if all t_k are positive, then

$$\Big\|\sum_{k\in\mu} t_k y_k\Big\| \geqq \Big|\beta\Big(\sum_{k\in\mu} t_k y_k\Big)\Big| = \sum_{k\in\mu} t_k \beta(y_k) \geqq \varepsilon \sum_{k\in\mu} t_k .$$

But if the t_k are of arbitrary sign, this with Theorem 1, (v), shows that

$$\Big\|\sum_{k\in\mu} t_k y_k\Big\| = \Big\|\sum_{k\in\mu} |t_k|\, y_k\Big\| \geqq \varepsilon \sum_{k\in\mu} |t_k| .$$

Also

$$\Big\|\sum_{k\in\mu} t_k y_k\Big\| \leqq \sum_{k\in\mu} |t_k|\, \|y_k\| = \sum_{k\in\mu} |t_k| .$$

It follows that $T(t_k, k\in\omega)=\sum_{k\in\omega} t_k y_k$ determines an isomorphism of $l^1(\omega)$ with the closed subspace in B spanned by the y_k.

If (b_i) is a shrinking basis, (β_i) is a basis for B^*. Hence there can be no isomorphism of $l^1(\omega)$ into B, for B^*, a separable space, cannot have the non-separable space $l^1(\omega)^*$ as a factor space.

With these improvements on the results of James [1], we can now improve on a theorem of Karlin [2].

Theorem 4. *Let B be a Banach space with an unconditional basis; then the following conditions are equivalent:*

(i) *B is reflexive.*

(ii) *B is w-ω-complete and contains no subspace isomorphic to $l^1(\omega)$.*

(iii) *B contains no subspace isomorphic to $c_0(\omega)$ or $l^1(\omega)$.*

(iv) *Neither B nor B^* contains a subspace isomorphic to $l^1(\omega)$.*

(v) *B^{**} is separable.*

Proof. By Theorems 2 and 3, and 3,3 the conditions (i), (ii), and (iii) are equivalent. (i) implies (v) for B is separable. (v) implies (iv) because $l^1(\omega)^*$ and $l^1(\omega)^{**}$ are both non-separable, so neither can be isomorphic to a subspace or factor space of B^{**}. To see that (iv) implies (iii) one need only observe that if T is an isomorphism of $c_0(\omega)$ into B, then T^* is a homomorphism of B^* onto $c_0(\omega)^*$, which is isomorphic to $l^1(\omega)$. By II, 2, (12), $l^1(\omega)$ is isomorphic to a subspace of B^*, contradicting part of (iii).

Theorem 5. *If* (b_i) *is an unconditional basis for* B, *and if* (β_i) *is the corresponding sequence in* B^*, *then the following conditions are equivalent;*

(i) (b_i) *is a shrinking basis for* B.

(ii) *No subspace of* B *is isomorphic to* $l^1(\omega)$.

(iii) (β_i) *is a basis for* B^*.

(iv) (β_i) *is a boundedly complete basis for* B^*.

(v) B^* *is separable.*

(vi) (β_i) *is an unconditional basis for* B^*.

(vii) B^* *is weakly sequentially complete.*

Proof. Equivalence of the conditions (i) to (v) and the implications "(vi)⇒(v)", and "(iv) and (vi) imply (vii)" have been established before. To see that (iii) implies (vi), renorm if necessary to get $\|V_\mu\| \leqq 1$ for all μ in $\sum$ and take β in B^*. For each $\varepsilon > 0$ there is $m = m_\varepsilon$ such that $\|V_n^* \beta\| < \varepsilon$ if $n > m$. But if $\mu \in \Sigma$ and μ contains all integers $\leqq m$, then $V_m V_\mu = V_\mu$ and $\|V_\mu^* \beta\| = \|V_\mu^* V_m^* \beta\| < \varepsilon$, so $\lim_{\mu \in \Sigma} \|V_\mu^* \beta\| = 0$; that is, (vi) holds.

To prove (vii)⇒(iii), for each β in B^*, $\|U_\mu^* \beta\| \leqq \|\beta\|$ if $\|\ldots\|$ has the properties of Theorem 1, (v). Hence the partial sums $\sum_{i \in \mu} \beta(b_i)\beta_i$ are uniformly bounded; by § 1, (4e), $\sum_{i \in \omega} \beta(b_i)\beta_i$ is w-absolutely convergent, hence w-subseries convergent. By the Orlicz-Pettis theorem 1, 1, $\sum_{i \in \omega} \beta(b_i)\beta_i$ is norm unconditionally convergent. But the series is w^*-unconditionally convergent to β, so the sum in the norm topology can be nothing but β, and (vii) implies (iii).

While it was still not determined whether or not a basis existed in every separable space, Karlin [2] observed that this result settles in the negative the corresponding question for unconditional bases.

Corollary 1. $C[0,1]$ *has no unconditional basis.*

Proof. $C[0,1]^*$ is an (AL)-space (Theorem VI, 1, 2) and therefore (Theorem VI, 4,4) is w-sequentially complete. But $C[0,1]^*$ is non-separable. By Theorem 5, $C[0,1]$ can have no unconditional basis.

(1) The notion of absolute basis (Definition 1) is not very fruitful. If (b_i) is such a basis for B with every $\|b_i\| = 1$, define T from $l^1(\omega)$ into

B by $T(\eta_i)=\sum_{i\in\omega}\eta_i b_i$. Then T is one-to-one and onto; by the interior mapping theorem II, 3,4, T is an isomorphism of B with $l^1(\omega)$.

(2) Generalizations of bases of many sorts have been studied. The book of Singer should be consulted for exhaustive information on bases. (a) In every separable normed space there exists a biorthogonal system (b_i), (β_i) for which (b_i) is fundamental in N and (β_i) total over N [Markusevic]. Such systems can be derived by extending similar systems in closed subspaces. (See V. D. Mil'man [10] Chapter 1, § 3, for a discussion of this.) (b) If there exist continuous linear projections $(P_i, i\in\omega)$ for which $P_i P_j=\delta_{ij} P_i$ for all i, j in ω, and if $\sum_{i\in\omega} P_i x$ converges to x for each x in B, then the range spaces $P_i(B)$, $i\in\omega$, make up a *Schauder basis of subspaces* of B. For some results of this sort see Retherford for references, Grinblyum for first appearance. (c) If for some uncountable index set S there is in B a biorthogonal system $(b_s, \beta_s, s\in S)$ such that for each x in B, $\sum_{s\in S}\beta_s(x)b_s$ converges (in the sense of Defn. II, 2,1) to x, Marti calls (b_s) a generalized basis and proves many of the usual properties of unconditional bases for these larger systems. Note that in $c_0(S)$ or $l_p(S)$ with $1\le p<\infty$, (δ_s) forms such a generalized basis. (d) The similarities between $l_p(S)$ and $L^p(S,\mu)$ with μ a finite measure suggests that the family of projections P_σ, σ a finite subset of S, might be replaced in the analogue by a larger bounded Boolean algebra of projections. Lindenstrauss and Zippin have devoted some attention to spaces possessing a sufficiently large bounded Boolean algebra of projections; in many ways their properties resemble those of spaces with bases. If the projections have finite-dimensional ranges this condition is stronger than the bounded approximation property.

Chapter V. Compact Convex Sets and Continuous Function Spaces

§ 1. Extreme Points of Compact Convex Sets

The present section is concerned with a cycle of theorems in all of which D. P. Mil'man had a part. They were originally stated for the w^*-topology of a conjugate space, but the proofs adapt without difficulty to locally convex spaces.

Definition 1. Let K be an convex subset of a linear space L. A point x of K is a *passing point* of K if x belongs to an open segment which is contained in K. A point of K which is not a passing point of K is an *extreme point* of K.

(1) x is an extreme point of K if and only if (a) $x=(x_1+x_2)/2$ and $x_i \in K$ imply that $x=x_1=x_2$, and if and only if (b) whenever x is a convex combination of a finite set φ of points of K, then x is an element of φ.

The general form of the Kreĭn-Mil'man theorem is

Theorem 1. *Let K be a compact, convex subset of an* LCS L *and let E be the set of extreme points of K. Then K is $K(E)$, the closed convex hull of E.*

To prove this and Theorem 2 we prove first only the partial result.

(A) Every (even weakly) compact, convex subset K of an LTS L for which L^* is total over L has at least one extreme point.

Proof of (A). Call a set A *extremal* if (i) A is a closed, convex non-empty subset of K, and (ii) every open segment S in K which contains a point of A lies wholly in A. Order the family $\mathfrak{A}$ of extremal subsets by $\subseteqq$; then $K \in \mathfrak{A}$ so $\mathfrak{A}$ is non-empty. If $\mathfrak{A}'$ is a linearly ordered subfamily and A_0 is the intersection of its elements, then $A_0 \in \mathfrak{A}$, so $\mathfrak{A}$ satisfies the hypotheses of Zorn's lemma, and *there is a minimal A in $\mathfrak{A}$*.

If an extremal A contains a passing point p of K, there is an open segment $S \subseteqq A$ such that p is in S. Since L^* is total over L, there exists an f in L^* such that f is not constant on S. Let $c=\sup\{f(x): x \in A\}$, and let $A_1=\{x: x \in A \text{ and } f(x)=c\}$. Then A_1 is a closed, convex, proper subset of A; by compactness of A, A_1 is not empty. To show that A_1

satisfies (ii) take an open segment S_1 in K which contains a point p_1 of A_1. Then S_1 is contained in A, as A satisfies (ii). Hence $f(x) \leqq c$ if $x \in S_1$, but $f(p_1)=c$, so $f(x)=c$ for all x in S_1; hence $S_1 \subseteqq A_1$.

This shows that a minimal A of $\mathfrak{A}$ cannot have in it a passing point of K; hence it contains only one point and that an extreme point of K.

The rest of the Kreĭn-Mil'man theorem can be derived from (A) and Mazur's theorem (Cor. I, 6,2); instead, we shall derive it from (A) and Theorem 2, which is a generalization of a theorem of Mil'man and Rutman whose proof uses Mazur's theorem in about the same way.

(2) Let K be a convex, compact set in an LCS L; define p in L^* by $p(f)=\sup f(K)$ for all f in L^*. (a) p is sublinear. (b) p is continuous in the $\mathfrak{K}$-topology of L^* (see I, 5, (2)); that is in the topology of uniform convergence on compact convex sets. (c) Hence by I, 5, (5b), an element ξ of $L^{*\#}$ which is dominated by p is in $Q(L)$. (d) If W is a wedge in L^* such that p is additive in W, then by the Hahn-Banach theorem there is a ξ in $L^{*\#}$ such that $\xi(f) \leqq p(f)$ for all f in L^* and $\xi(f)=p(f)$ for all f in W. (e) By (c), (d), and Mazur's theorem I, 6,3 there is an x in K with $\xi=Qx$, so $p(f) \geqq f(x)$ for all f in L^* and $p(f)=f(x)$ if $f \in W$.

Lemma 1. *Let K be a convex, compact subset of an* LCS *L, let $p(f)=\sup f(K)$ for each f in L^*, and let $\mathfrak{W}$ be the family of all wedges W in L^* such that p is additive in W; order $\mathfrak{W}$ by $\supseteqq$. Then every element f of L^*, and every element W' of $\mathfrak{W}$, is contained in a maximal W of $\mathfrak{W}$.*

Proof. $\mathfrak{W}$ contains $\{0\}$ and every half-ray $\{tf: t \geqq 0\}$. Also the union of the elements of any simply ordered subfamily of $\mathfrak{W}$ is again an element of $\mathfrak{W}$; Zorn's lemma gives the desired conclusion.

(3) Given L, K, and $\mathfrak{W}$ as in Lemma 1, for each x in K define $W(x)=\{f: f(x)=p(f)\}$. For each [maximal] W in $\mathfrak{W}$ define F_W, the [*minimal*] *facet* of K determined by W, to be $\{x: W(x) \supseteqq W\}$. Then: (a) Each $W(x) \in \mathfrak{W}$; there may be x in K such that $W(x)=\{0\}$. $K=F_{\{0\}}$. (b) (2e) asserts that every $W \subseteqq$ some $W(x)$; hence every F_W is non-empty. (c) Every F_W is an extremal subset of K, so every extreme point of F_W is an extreme point K. (d) $W' \subseteqq W$ implies $F_{W'} \supseteqq F_W$; hence every facet contains a minimal facet. (e) If $W_i \in \mathfrak{W}$, then $W_1+W_2 \in \mathfrak{W}$ if and only if $F_{W_1} \cap F_{W_2} \neq 0$. (f) Hence distinct minimal facets of K are disjoint, and if F_W has a point in common with a minimal facet $F_{W'}$, then $F_{W'} \subseteqq F_W$.

(4) As an example, let L be the plane and let K be the convex hull of two equal tangent circles. Then the set $\{0\}$ is the $W(x)$ for every interior point x of K. The rest of $\mathfrak{W}$ is the set of half-rays in L^* radiating from the origin. The minimal facets of K are the single points of the open semicircles of the boundary of K and the two closed line segments of the boundary of K.

Theorem 2. *Let K be a compact, convex subset of an* LCS *L and let G be a subset of K containing at least one point from each minimal facet of K. Then $K=K(G)$, the closed convex hull of G.*

Clearly $K \supseteqq K(G)$. By Mazur's theorem (Cor. 1,6,2)

$$K(G)=\bigcap_{f\in L^*}\{x: f(x)\leqq \sup f(G)\}.$$

But if $f\in L^*$, there is a maximal W in $\mathfrak{W}$ in which f lies; by hypothesis there is a point y in $F_W \cap G$; then $f\in W(y)$ and if $x\in K$, then

$$f(x)\leqq f(y)\leqq \sup f(G).$$

Hence $K\subseteqq K(G)$.

This proves Theorem 2. To derive Theorem 1 from this, note that (A) and (3c) imply that E, the set of extreme points of K has the properties required of G in Theorem 2.

The Kreĭn-Mil'man theorem has a partial converse, Theorem 3 below, due to Mil'man [1]. In its proof we use a lemma of Bourbaki [1, p. 80].

Lemma 2. *If $K_i, i=1,\ldots,n$, are compact, convex subsets of an* LCS *L, then the convex hull of $\bigcup_{i\leqq n} K_i$ is compact.*

Proof. Let I be the closed interval $[0,1]$. In $L\times R$ each set $K_i\times I$ is compact and convex; by Tyhonov's theorem $K=\prod_{i\leqq n}(K_i\times I)$ is a compact convex subset of $(L\times R)^n$. The set S of all $(x_1,t_1,\ldots,x_n,t_n)$ in $(L\times R)^n$ for which $\sum_{i\leqq n} t_i=1$ is closed, so $S\cap K$ is compact. The function F carrying $(x_1,t_1,\ldots,x_n,t_n)$ to $\sum_{i\leqq n} t_i x_i$ is continuous and $F(S\cap K)$ is the convex hull of the union of the K_i; hence the convex hull is compact. (Kelley, p. 141.)

Theorem 3. *Let L be an* LCS *and either* (i) *let L be topologically complete* (Def. III, 1,1) *and let A be a totally bounded subset of L or* (ii) *let A be a subset of a compact, convex subset of L. Then the closure of A contains the set E of extreme points of $K(A)$, the closed convex hull of A.*

Proof. Under either hypothesis $K(A)$ is compact so E is non-empty. Let x be a point of E and let V be a closed, convex, symmetric neighborhood of 0 in L; by total boundedness of A there exist $y_1,\ldots,y_n$ in A such that $\bigcup_{i\leqq n}(y_i+V)\supseteqq A$. Let $K_i=$closed convex hull of $A\cap(y_i+V)$. Then $K_i\subseteqq K(A)\cap(y_i+V)$ so K_i is compact; clearly $K\left(\bigcup_{i\leqq n} K_i\right)=K(A)$. But $K\left(\bigcup_{i\leqq n} K_i\right)$ is just the convex hull of the K_i, for, by Lemma 2, the convex hull of n compact, convex sets is again compact. Hence $x=\sum_{i\leqq n} t_i(y_i+v_i)$, where $v_i\in V$, $t_i\geqq 0$ and $\sum_{i\leqq n} t_i=1$, and $y_i+v_i\in K_i\subseteqq K(A)$. By (1b) there is an i such that $x=y_i+v_i\in A+V$, so $(x+V)\cap A\neq\emptyset$. This holds for every V in a neighborhood basis at 0; that is, x is in the closure of A.

These can all be collected into one theorem, basically formulated by Klee [4], to be called here the

***K*²-*M*³-*R* Theorem.** *Let K be a convex, compact set in an* LCS *L and let E be the set of extreme points of K. Then the following conditions on a subset A of K are equivalent:*

(i) *$K(A)$, the closed convex hull of A, is K.*

(ii) *The closure of A contains E.*

(iii) *The closure of A contains at least one point of each minimal facet of K.*

(iv) *For each f in L^*, $\sup f(A)=\sup f(K)$.*

(5) The original paper of Kreĭn and Mil'man considered the case of a bounded w^*-closed convex set in N^*, the conjugate of a normed space. (a) Every such convex set is the w^*-closure if its set of extreme points; in particular, (b) the unit ball in N^* is the w^*-closed convex hull of its set of extreme points. (c) If B is a reflexive Banach space, the unit ball is the norm-closed convex hull of its set of extreme points. (d) If the unit ball in N has only a finite number of extreme points but N is not finite-dimensional, then N is not isometric to a conjugate space.

(6) Examples pertinent to (5) are: (a) It happens that the unit ball of $l^1(\omega)$ is the norm-closed convex hull of its set of extreme points; however, from (5b) and the isometry of $l^1(\omega)$ with $c_0(\omega)^*$ it is only possible to conclude that the unit ball of $l^1(\omega)$ is the closure under coordinatewise convergence of the convex hull of its set of extreme points. (b) In $l^1(\omega)$ each extreme point of the unit ball is of the form $\pm\delta_j$, where δ_j is the sequence with 1 in the j'th place and 0 elsewhere. (c) $C[0,1]$ is not isometric to any conjugate space, for there are only two extreme points of the unit ball, the functions of absolute value 1. (d) The unit ball of $c_0(S)$ has no extreme points, so $c_0(S)$ is not isometric to any conjugate space. (e) $L^1[0,1]$ is not isometric to any conjugate space, for the same reason. A modification of the proof of (b) shows that if μ is a measure, then $\pm x$ is an extreme point of the unit ball in $L^1(\mu)$ if and only if there is an atom A with respect to μ (that is, a μ-measurable set A such that $0<\mu(A)<\infty$ and such that if A' is a μ-measurable subset of A then $\mu(A')=\mu(A)$ or $\mu(A')=0$) such that x is almost everywhere equal to the characteristic function of A divided by $\mu(A)$. (f) For every measure μ and every p with $1<p<\infty$, every point of norm 1 is an extreme point of the unit ball of $L^p(\mu)$.

Gel'fand showed that $L^1[0,1]$ is not even isomorphic to a conjugate space.

(7) Let D be a bounded region in the complex plane, let E be its closure, and let A be the Banach space of all continuous complex functions on E which are analytic in D. (a) Šilov (see Gel'fand, Raĭkov and Šilov, § 24) discussed the set $B \subseteq E$ which is now called the *Šilov boundary of* E; B is the smallest closed subset of E such that every element of A attains its maximum absolute value at some point of B. For example, if D is a disc with the origin omitted, $D=\{z: 0<|z|<1\}$, then $B=B(E)=\{z: |z|=1\}$. (b) Mil'man [1] pointed out that if K is the w^*-closed convex hull in A^* of the set of evaluation functionals $\{\varphi_s: s \in E\}$ (where $\varphi_s(x)=x(s)$ for every x in A), then s is in the Šilov boundary of E if and only if φ_s is in the w^*-closure of the set of extreme points of K.

(8) There are several properties of a corner of a polygon which do not remain equivalent when carried to more general convex sets K. (i) a is an *extreme point* of K means that a is not the midpoint of any segment of positive length contained in K. (ii) a is a *denting point* for K if for each $\varepsilon>0$, a is not in the closed convex hull of $K \setminus (\varepsilon$-ball about $a)$. (iii) a is a *vertex* of K if the set of f which attain their supremum on K at a is total over L. (iv) a is a *strongly extreme point* of K in N if for each $\varepsilon>0$ there is $\delta=\delta(\varepsilon)>0$ such that a is not the middle of any segment of length ε which is contained in $K+\delta U$. (v) a is an *exposed point* of K *(exposed by* α*)* if there is α in L^* such that $\alpha(a)=\sup \alpha(K)>\alpha(y)$ if $a \neq y \in K$. (vi) In N, a is a *strongly exposed point (by* α*) of* K if for each $\varepsilon>0$ there is $\delta(\varepsilon)>0$ such that for each x in K, $\alpha(x) \leq \alpha(a)-\delta(\|x-a\|)$. (vii) A convex body K is *uniformly rotund from a boundary point* a if for each $\varepsilon>0$ there is $\delta(a,\varepsilon)>0$ such that if y is in K and $\|y-a\| \geq \varepsilon$, then $(a+y)/2$ is at least $\delta(a,\varepsilon)$ away from the complement of K. (This one is not a property of any boundary point of a polygon but it fits in here.)

$$\begin{array}{lllll} \text{(a)}\ \ \text{(vii)} \Rightarrow & \text{(vi)} \Rightarrow & \text{(iv)} \Leftrightarrow & \text{(ii)} \\ & \Downarrow & \Downarrow & \\ \text{(iii)} \Rightarrow & \text{(v)} \Rightarrow & \text{(i)}. & \end{array}$$

(b) Straszewicz and later Mil'man [2] discussed exposed points (called accessible points in the review in MR 9,449.) Klee [4] points out that extreme points of K do not depend on the topology of L, but exposed points depend on L^*. These two and vertices can be studied as well in an LCS, but the others mention a norm in the definition. However they are not affected by shifting to an isomorphic norm. (c) Bohnenblust and Karlin proved that the identity element of a Banach algebra is a vertex of the unit ball. (d) Mil'man [2] shows that if S is compact metric and L is a linear subspace of $C(S)$ which separates points of S, and if K is the closed convex hull in L^* of the set of evaluation functionals of S, $\varphi(S)$, then the set of w^*-exposed points of K is w^*-dense in the Silov boundary $w^*(\varphi(S))$. [See also Dales.] (e) By III, 5, (15a) each w-compact convex K in a Banach space is the closed convex hull of its set of exposed points. (f) If B is (UR) or (LUR) as defined in VII, § 2, then every point of the unit sphere satisfies (vii).

(9) (a) Lindenstrauss [3] shows that *if K is convex and w-compact in an* (LUR) *space, then K is the closed convex hull of its strongly exposed points.* The things we have carried through in this book will not give quite this much, only (8e). (b) Theorem VII, 4, 1 shows that each reflexive or (WCG) space can be renormed to be (LUR). Since strong exposure is unaffected by isomorphism, we conclude from applying Th. VII, 4, (1a) to the space spanned by K that *in every normed space every convex, w-compact set is the closed convex hull of its set of strongly exposed points.* This with (8a), (vi)⇒(ii) settles the Question 1 of Rieffel in the affirmative.

(10) (a) Asplund and Namioka need for their proof of the Ryll-Nardzewski fixed point theorem the following lemma, which in case there is a *countable* dense subset in K, they prove by direct calculation. *If K is a convex, w-compact subset of an* LCS *L, then for each continuous prenorm p on L and each* $d>0$ *there is a closed convex subset C of K such that* $C \neq K$ *and the p-diameter of* $K \setminus C$ *is* $<d$. (b) This reduces at once to the Banach space case by passing to B, the completion of $L/p^{-1}(0)$; there it is equivalent to the following consequence of (9b): *Every w-compact, convex K in B has a denting point*, that is, *a strongly extreme point.* (c) The direct proof of (a) uses a result of Choquet [Dixmier, 4, p. 355, B 14.]

The w-topologized set of extreme points of a compact convex K in a separable normed space N is a space of second category in itself.

§ 2. Fixed-point Theorems

The basic theorem is well known in the finite-dimensional case; it is extremely useful in existence proofs for linear functionals fixed under certain groups of linear transformations. We assume the fundamental fixed point theorem of Brouwer as our starting point; a number of proofs have been given including a very brief one by Knaster, Kuratowski, and Mazurkiewicz.

Brouwer fixed-point theorem. *If S is homeomorphic to a closed n-cell,* $\{x\colon \|x\| \leqq 1\}$, *in a Euclidean n-space and if F is a continuous function mapping S into itself, then there is a point p of S fixed under F; that is, a p in S with* $F(p)=p$.

Schauder [2] extended this theorem to compact convex subsets of normed spaces; his proof is easily adapted to locally convex spaces [Tyhonov 2].

Theorem 1. *Let K be a convex compact set in a locally convex linear topological space L and let F be a continuous function from K into itself; then F has a fixed point in K.*

Proof. Continuity of F and compactness of K imply [Kelley, p. 198] that F is uniformly continuous on K; that is, for each convex, symmetric neighborhood U of 0 in L there exists a convex, symmetric neighborhood V of 0 in L such that $V \subseteqq U$ and such that $F(x)-F(x') \in U$ if $x, x' \in K$ and $x - x' \in V$.

For such a U we shall construct an approximating function F_U on a convex, closed, finite-dimensional subset K_U of K. Let $x_1, \ldots, x_n$ be a finite subset of K such that $K \subseteqq \bigcup_{i \leqq n} (x_i + V)$. Then K_U, the convex hull of this finite set, is isomorphic to a polyhedron in some Euclidean space of dimension $\leqq n$; hence K_U can be subdivided into disjoint open simplexes $\sigma_1, \ldots, \sigma_k$ such that each σ_i is countained in a translation of V. If y is a vertex of a σ_i, define $F_U(y)$ so that $F(y) - F_U(y) \in V$ and $F_U(y) \in K_U$; then if $x \in K_U$ and $y_1, \ldots, y_m$ are the vertices of the simplex containing x, so that $x = \sum_{j \leqq m} t_j y_j$, set $F_U(x) = \sum_{j \leqq m} t_j F_U(y_j)$.

Then F_U is continuous and can be shown to approximate F within $3U$ everywhere in K_U. K_U and F_U satisfy the conditions of Brouwer's theorem, so there is a point p_U in K_U such that $F_U(p_U) = p_U$. Let $\mathfrak{U}$ be the family of convex, symmetric neighborhoods of 0, directed by $\subseteqq$. By compactness of K, the net $(p_U, U \in \mathfrak{U})$ has a subnet $(p_d, d \in \Delta)$ which converges to a point p of K. It can be verified that $F(p) = p$.

Corollary 1. *If F maps a topological space X into a subset E which is homeomorphic to a compact convex subset of an* LCS *L, then F has a fixed point in E.*

Corollary 2. *Let L be an* LCS *in which the closed convex hull of a compact set is compact and let F be a continuous function carrying some convex subset K of L into a compact subset K′ of K. Then F has a fixed point in K.*

[The hypothesis of this corollary is valid in a complete LCS, or in what von Neumann called a topologically complete space, Def. III, 1, 1.]

As an illustration of the applications of this theorem we prove here the existence of "Banach limits" for the space $m(\omega)$.

Definition 1. If Σ is a set, in each linear subspace X of $m(\Sigma)$ which contains the constantly 1 function e, a *mean* μ is an element of $m(\Sigma)^*$ such that for each x in X, $\inf x(\Sigma) \leq \mu(x) \leq \sup x(\Sigma)$.

(1) The set M of means on X is the w^*-closed convex hull of the set of evaluation functionals $\{\delta_\sigma : \delta_\sigma(x) = x(\sigma)$ for all x in $X\}$. (See Thms. 1,1 and 3.)

(2) If K is a compact convex set in an LCS L, let $A(K)$ be the set of affine, continuous functions from K into R. (a) Then each mean v on $A(K)$ can be represented by evaluation by a (unique) point $\pi(v)$ in K. [By (1) there exist convex combinations $v_\sigma = \sum_{k \in \sigma} t_{\sigma k} \delta_k$ such that $v = w^*\text{-}\lim v_\sigma$. Let $p_\sigma = \sum_{k \in \sigma} t_{\sigma k} k$ and take a subnet $(p_{\sigma(n)}, n \in \Delta)$ which converges to some point $\pi(v)$ in K. Then for all x in $A(K)$, $x(\pi(v)) = \lim_n x(p_{\sigma(n)}) = \lim_n v_{\sigma(n)}(x) = v(x)$. $\pi(v)$ is unique because $L^*_{|K} \subseteqq A(K)$ separates points of K.] (b) Hence *in* $A(K)$ *weak convergence is the same as coordinatewise convergence.* (c) In $A(K)$ the vector sum of $L^*_{|K}$ and the constant functions is norm dense in $A(K)$. [Phelps, p. 4.]

Definition 2. If an associative, binary operation $(\sigma, \tau) \to \sigma\tau$ is defined in Σ (that is, if Σ is a semigroup), the *left* [*right*] *shift* by σ in $m(\Sigma)$, $l_\sigma [r_\sigma]$, is defined for each x in $m(\Sigma)$ by $[l_\sigma x](\tau) = x(\sigma\tau)$ $[[r_\sigma x](\tau) = x(\tau\sigma)]$ for all τ in Σ. Then μ is a *left* [*right*] *invariant mean on* X if for each σ in Σ, $l_\sigma(X) \subseteqq X [r_\sigma(X) \subseteqq X]$ and $l_\sigma^* \mu = \mu [r_\sigma^* \mu = \mu]$. Σ is *left* [*right*] *amenable* when there is at least one left [right] invariant mean on $m(\Sigma)$. (See Day [14] and [18] and Greenleaf for the history, theory, and applications of amenability of semigroups and groups.)

Theorem 2. *There is an invariant mean* μ *in* $m(\omega)^*$*; for any such* μ, *also* (i) $\|\mu\| = 1$, *and* (ii) $\liminf_{n \in \omega} x_n \leq \mu(x) \leq \limsup_{n \in \omega} x_n$ *for each* x *in* $m(\Sigma)$.

Proof. The set M of means on $m(\Sigma)$ is the intersection of U^π with the hyperplane $\{v : v(e) = 1\}$. Hence M is convex and w^*-compact and if $T = l_1$, then T^* is w^*-w^*-continuous by Th. I, 5,2, and $T^*(M) \subseteqq M$. By Th. 1, T^* has a fixed point μ which is clearly fixed under all $(l_n)^* = (T^*)^n$. (i) is clear for this μ. To prove (ii) note first that in $m(\omega)$ the conditions $\|\mu\| = \mu(e) = 1$ imply that $\mu(x) \geqq 0$ if all $x_n \geqq 0$. [For otherwise

$$\mu\left(e - \frac{x}{\|x\|}\right) = \mu(e) - \frac{\mu(x)}{\|x\|} > \mu(e) = 1 = \|\mu\|,$$

while $\|e - x/\|x\|\,\| \leqq 1$.] Also, if x is in the subspace $c_0(\omega)$ of $m(\omega)$, then $\|T^n x\| \to 0$, so $\mu(T^n x) \to 0$. By invariance $\mu(x) = 0$ if $x \in c_0(\omega)$. Now if $y \in m(\omega)$, there exists x in $c_0(\omega)$ such that

$$\limsup_{n \in \omega} y_n \geqq \sup_{n \in \omega} (x_n + y_n) \geqq \inf_{n \in \omega} (x_n + y_n) \geqq \liminf_{n \in \omega} y_n.$$

Then $\mu(y)=\mu(x+y)$ and

$$\limsup_{n\in\omega} y_n=\mu\left(e\limsup_{n\in\omega} y_n\right)\geqq\mu(x+y)=\mu(y)\geqq\liminf_{n\in\omega} y_n.$$

Theorem 3 [Markov]. *Let K be a compact, convex subset of an* LCS *L and let $(F_s, s\in S)$ be a commuting family of continuous functions from K into itself such that each F_s is affine; that is, for x, y in K and $0<t<1$, $F_s(tx+(1-t)y)=tF_s(x)+(1-t)F_s(y)$. Then the F_s have a common fixed point in K.*

Proof. Let K_s be the set of fixed points of F_s. Theorem 1 asserts that each K_s is non-empty; affineness implies that K_s is convex, continuity that K_s is closed; hence each K_s shares the properties of K including compactness. But commutativity of F_s and F_t, s, t in S, imply that $F_s(K_t)\subseteqq K_t$. By Theorem 1, F_s has a fixed point in K_t, so $K_s\cap K_t$ again shares the properties assumed for K. By induction on the number of elements, $\bigcap_{s\in\sigma} K_s$ is non-empty for each finite subset σ of S. By compactness, $K'=\bigcap_{s\in S} K_s\neq\emptyset$; any element of K' is a fixed point common to all the F_s.

There is a simple direct proof of Theorem 3, probably due to Bourbaki, but certainly inspired by F. Riesz's proof of the ergodic theorem [1]. For σ in Σ, the set of finite subsets of S, and n in ω define

$$F_{\sigma n}=\prod_{s\in\sigma}\left[\frac{\left(\sum_{i\leqq n} F_s^i\right)}{n}\right].$$

By compactness of K, for each x in K the net $(F_{\sigma n}x, (\sigma,n)\in\Sigma\times\omega)$ has a convergent subnet $(F_{\sigma_m n_m}x, m\in\Delta)$ with some limit y. Then y is a fixed point of every F_s, for, if $s\in\sigma_m$ and $\tau_m=\sigma_m\setminus\{s\}$,

$$F_s y-y=\lim_{m\in\Delta} F_{\tau_m n_m}\cdot\frac{[F_s^{n_m+1}x-F_s x]}{n_m}=0,$$

because the expression after the limit is in $(K-K)/n_m$; by I, 4, (17b), this tends to 0 as $n_m\to\infty$.

Banach p. 33 proved that the additive group R is amenable by careful construction of a sublinear p to use in the Hahn-Banach theorem. Day [12] first adapted that proof for the following result.

Theorem 4. *Every abelian semigroup is amenable (left and right).*

Proof. Apply Th. 3 to the set M of means on Σ (as in the proof of Th. 2) and the family $\{l_\sigma^*:\sigma\in\Sigma\}$ restricted to M.

Definition 3. If K is a convex compact subset of an LCS L, let $\mathscr{A}(K)$ be the semigroup of affine continuous transformations of K into K, with composition for multiplication.

Theorem 5 [Day 17]. *A semigroup Σ is left amenable if and only if for each compact convex subset K of each* LCS *L and for each homomorphism h of Σ into $\mathscr{A}(K)$ there is in K a common fixed point p of all the transformations h_σ.*

Remark. There is also a version of this for topological semigroups; see Day [18], § 6.

Proof. Sufficiency is simple because the choice $K=$ set M of means on $m(\Sigma)$, $h_\sigma = l^*_{\sigma|K}$ describes a homomorphism of Σ into $\mathscr{A}(M)$ whose fixed points are the left-invariant means. For the converse, choose y arbitrarily in K and define T from $L^*_{|K} \subseteqq A(K)$ by $[T\varphi](\sigma) = \varphi(h_\sigma y)$ for all σ in Σ. Then T^* carries means on $m(\Sigma)$ to means on $A(K)$, and, in particular, $\pi T^* \delta_\tau = h_\tau y$ for each τ in Σ. If μ is a left-invariant mean on $m(\Sigma)$, then $\nu = T^* \mu$ is a mean on $A(K)$ and setting $p = \pi(\nu)$ gives a point of K such that $\nu(\varphi) = \varphi(p)$ for all φ in $A(K)$. By (1), there are finite averages $\varphi_n = \sum_\tau a_{n\tau} \delta_\tau$ such that $\mu = w^*\text{-}\lim_{n\in\Delta} \varphi_n$. We can calculate that $l^*_\sigma \delta_\tau = \delta_{\sigma\tau}$ so

$$w^*\text{-}\lim_n \sum a_{n\tau} \delta_{\sigma\tau} = l^*_\sigma \mu = \mu = w^*\text{-}\lim_n \sum a_{n\tau} \delta_\tau .$$

Then

$$w^*\text{-}\lim_n T^*\Big(\sum a_{n\tau} \delta_{\sigma\tau}\Big) = T^* l^*_\sigma \mu = T^* \mu = w^*\text{-}\lim_n T^* \sum a_{n\tau} \delta_\tau ,$$

or for each φ in L^*,

$$\lim_n \varphi\Big(\sum a_{n\tau} h_{\sigma\tau} y\Big) = \varphi(\pi T^* l^*_\sigma \mu) = \varphi(\pi T^* \mu) = \varphi(p)$$

but the first quantity also equals

$$\lim_n \varphi\Big(h_\sigma\Big(\sum a_{n\tau} h_\tau\Big)\Big) = \lim_n [\varphi \circ h_\sigma]\Big(\sum a_{n\tau} h_\tau\Big) = [\varphi \circ h_\sigma](p) = \varphi(h_\sigma p),$$

so p is invariant under all h_σ, because L^* separates points of K.

Theorems 2 and 4 imposed some structure conditions on the semigroups of affine continuous mappings. Other theorems impose some more geometric conditions on K or Σ. The next important fixed point theorem is due to Ryll-Nardzewski; the present proof is from Asplund and Namioka.

Definition 4. A semigroup $\mathscr{S}$ of transformations of a subset E of an LCS L is called *distal* (sometimes *noncontracting*) if whenever $x \neq y$, then 0 is not in the closure of $\{Sx - Sy : S \in \mathscr{S}\}$.

(3) In an LCS L a semigroup $\mathscr{S}$ is distal if and only if there is a prenorm p and a number $d > 0$ such that $p(Sx - Sy) > d$ for all S in $\mathscr{S}$.

Theorem 6 (Ryll-Nardzewski). *Let K be a convex, w-compact subset of an* LCS *L. Then each distal semigroup $\mathscr{S}$ of w-continuous affine maps of K into itself has a common fixed point p in K.*

Proof. It suffices to prove that each finite subset $T_1, \ldots, T_n$ of $\mathscr{S}$ has a common fixed point, so we begin with the case where the T_i generate $\mathscr{S}$, and let $T=(T_1+\cdots+T_n)/n$. T has a fixed point a. If all $T_i a=a$, we are finished; if not, discard those i for which the equality holds, renumber, and assume we are now in the case where $T_i a \neq a$ for all $i \leq n$. As $\mathscr{S}$ is distal, there exist a prenorm p and $d>0$ such that

(i) $p(S\,T_i a - S a)>d$ for all S in $\mathscr{S}$ and $i \leq n$. Let W be the closed convex hull of the orbit $\{S a : S \in \mathscr{S}\}$. By 1, (9a), there is a convex $C \subseteqq W$ such that the p-diameter of $W \setminus C<d$. The Mil'man Th. 1,3 says that there is S_0 in $\mathscr{S}$ with $S_0 a$ in $W \setminus C$. But $S_0 a=\left(\sum_{i \leq n} S_0 T_i a\right) n$, so at least one $S_0 T_i a$ is in $W \setminus C$, a contradiction with (i), so $T_i a=a$ for all $i \leqq n$.

Compactness carries the rest of the proof, as in Th. 3.

Corollary 3 (Ryll-Nardzewski). *Let Σ be a group and let W be the family of all weakly almost periodic functions in $m(\Sigma)$; that is, x is in W means that $\{l_\sigma x : \sigma \in \Sigma\} \cup \{r_\sigma x : \sigma \in \Sigma\}$ is relatively w-compact in $m(\Sigma)$. Then there is a unique invariant mean on W.*

Proof. For x in W let L_x and R_x be the closed convex hulls of $\{l_\sigma x : \sigma \in \Sigma\}$ and $\{r_\sigma\, x : \sigma \in \Sigma\}$. Each l_σ and each r_σ is an isometry, so these groups are distal in L_x and R_x. By Theorem 5 there is a fixed element, $t e$, in L_x and another, $u e$, in R_x. By Mazur's theorem in the form of Th. II, 5,2, there exist nets of finite averages λ_n of the l_σ and ρ_n of the r_σ such that $\|\lambda_n x - t e\| \to 0$ and $\|\rho_m x - u e\| \to 0$. But λ_n and ρ_m commute (by associativity of Σ), so $\lambda_n \rho_m x$ tends to te and to $u e$. Therefore each invariant element of L_x equals each invariant element of $R_x = t e$. Setting $\mu(x)=t$ gives an invariant mean on W.

Let us consider next a still more geometric condition, this time applied to K directly.

(4) The *radius $r_p(A)$ of a bounded set A from a point p* is $\sup\{\|p-x\| : x \in A\}$. The *diameter* of A, diam A, is $\sup\{\|x-y\| : x \in A, y \in A\}$. If C is another set in N, then the *Čebyšev radius for A in C*, $r(A,C)$, $=\inf\{r_c(A) : c \in C\}$, and the *set of Čebyšev centers of A in C*, $\check{C}(A,C)$, $=\{c \in C : r_c(A)=r(A,C)\}$.

(a) If K is convex and w-compact, then the set of Čebyšev centers of K in K is not empty. [If $E_t=\{p \in K : r_p(K)<t\}$, then the Čebyšev radius of K, $r=r(K,K)$, $=\inf\{t : E_t \neq \emptyset\}$. Then each E_t, $t>r$, is the intersection of closed balls with K so is convex and w-compact; hence $E_r = \bigcap_{t>r} E_t$

is non-empty.] (b) A set $A \subseteqq N$ is said to have *normal structure* if for each closed convex bounded set W in A with more than one point there is a point p in W such that $r_p(W) < \operatorname{diam} W$. A convex, w-compact set K has normal structure if and only if for each closed convex $W \subseteqq K$, $\emptyset \neq \check{C}(W, W) \neq W$. [$\check{C}(W, W)$ is not empty by (a); $\check{C}(W, W) \neq W$ if and only if $r(W, W) < \operatorname{diam} W$.] (c) Brodskiĭ and Mil'man (who invented normal structure) prove that *if K is a convex, w-compact set with normal structure, then there is a common fixed point for the set of all isometries of K onto K.* [If $K_1 = \check{C}(K, K)$, (b) says that $K \neq K_1 \neq \emptyset$. K_1 is also convex and w-compact and is carried into itself by every isometry of K onto K. If K_1 has more than one point, repeat the same step, by transfinite induction if necessary, to get a fixed point.] (d) Belluce and Kirk [1] showed that if a w-compact, convex set K has normal structure, then every finite commuting set $T_1, \ldots, T_n$ of mappings of K into K which do not increase distance has a common fixed point in K. (e) Unfortunately the positive face M of the unit ball in an (AL) space does not fit this condition; if $p \in M$, then $r_p(M) = 2 = \operatorname{diam} M$. In particular, in $m(\Sigma)^*$ there may be no invariant mean even under the set of isometries $l_\sigma^*, \sigma \in \Sigma$; see Day [14], for example, for von Neumann's example; *the free group on more than one generator allows no invariant mean.* (f) Garkavi showed that for each bounded set A to have at most one Čebyšev center in N it is necessary and sufficient that N be uniformly rotund in every direction (VII, § 2). (g) Day, James and Swaminathan show that (Th. 5) if A is a non-empty bounded subset of a convex set S of a (UR_Σ) space, then $\check{C}(A, S)$ has at most one element.

(5) Belluce and Kirk [2] say that a closed convex set K in a Banach space has *completely normal structure* if for each bounded closed convex subset W of K with more than one point and and each decreasing net $(W_n, n \in \Delta)$ of subsets of W such that $r(W_n, W) = r(W, W)$ it follows that $W \neq$ the closure of the union of $\check{C}(W_n, W) \neq 0$. (a) If K has completely normal structure, it has normal structure. [Let $W_n = W$ for all n.] (b) Belluce and Kirk [2] introduced the concept of completely normal structure to generalize (4d). *If K is a w-compact convex set with completely normal structure, then every commuting family $\mathscr{S}$ of mappings of K into K which do not increase distances has a common fixed point.* (c) Belluce and Kirk [2] show that (a) includes a theorem of Browder (same as (a) but instead of assuming normal structure assume B is uniformly rotund) by showing that (UR) *implies that K has normal structure.* Browder also generalizes his theorem in later papers. (d) Day, James, and Swaminathan show that *every reflexive (UR_Σ) Banach space has completely normal structure.* [$\check{C}(W_n, W)$ cannot shrink as W_n decreases, but, by (4g), $\check{C}(W_n, W)$ has just one point, so that point is the same for all n.]

Banach used the Hahn-Banach theorem and a skillfully constructed sublinear p to prove the existence of a mean on $m(R)$ invariant under translation. For other applications of this technique see Agnew and also Silverman [1,2].

§ 3. Some Properties of Continuous Function Spaces

The space of continuous functions on a compact Hausdorff space has been studied from many points of view; the basic theorems of this section are due to M. H. Stone. These are the Stone-Čech compactification theory, the Banach-Stone theorem, and the Stone-Weierstrass theorem.

(1) If S is a topological space, let $C(S)$ be the subspace of all continuous functions in $m(S)$; then $C(S)$ and $C(S)^*$ are Banach spaces whose unit balls shall, as usual, be called U and U^π. Define q from S into $C(S)^*$ by $f_s = q(s)$ if $f_s(x) = x(s)$ for all x in $C(S)$. Then: (a) If $s \in S$, then $\|f_s\| = 1$. (b) Each f_s is an extreme point of the unit ball U^π in $C(S)^*$. (c) $q(s) = q(t)$ if and only if $x(s) = x(t)$ for all x in $C(S)$. (d) q is a continuous mapping of S into $C(S)^*$ in its w^*-topology. (e) q is a homeomorphism of S with $q(S)$ if and only if S is a *completely regular* [Tyhonov 1] space; that is, if and only if points in S are closed and for each s_0 in S and each neighborhood V of s_0 there is an x in $C(S)$ such that $x(s_0) \geqq 1$ and $x(s) \leqq 0$ if $s \notin V$.

(2) Let S be a completely regular space; then: (a) $q(S)$ (as defined in (1)) is closed if and only if S is compact. (b) U^π is the w^*-closed convex hull of $q(S) \cup -q(S)$. [by Mil'man's Theorem 1, 2.] (c) If $s \in S$ and $A(s) = \{x : x(s) = \|x\| = 1\}$, then $A(s)$ is a maximal convex subset of the surface of U. (d) If $W(s) = \{x : x(s) = \|x\|\}$, then $W(s)$ is a cone in $C(S)$ in which the norm is an additive function, and $W(s)$ is maximal under these conditions.

(3) Suppose that S is compact Hausdorff. (a) Then the set of extreme points of the unit ball U^π is closed (even equal to $q(S) \cup -q(S)$). (b) Every maximal convex subset of the surface of the unit ball U is of the form (2c) for some appropriate s in S, and every maximal cone in which the norm is additive is of the form (2d).

Using this collection of random facts we can prove the theorem of Stone [1] on compactification of a completely regular space S; Čech constructed, by a different method, a homeomorphic space with similar properties.

Definition 1. If S is a completely regular space and if q is the evaluation mapping of S into $C(S)^*$, let $\Omega(S)$, the *Stone-Čech compactification of* S, be the w^*-closure of $q(S)$ with its relative w^*-topology.

Theorem 1. *The Stone-Čech compactification of a completely regular space S has the following properties:* (i) *$\Omega(S)$ is a compact Hausdorff space and S is homeomorphic (under q) to a dense subspace of it.* (ii) *Hence S is homeomorphic to $\Omega(S)$ if and only if S is compact.* (iii) *If V is defined for all ξ in $C(\Omega(S))$ by $V\xi = \xi \circ q$, then V is an isometry of $C(\Omega(S))$ into $C(S)$.* (iv) *If Tx is defined by reducing the domain of definition of Qx to $\Omega(S)$, then T is an isometry of $C(S)$ into $C(\Omega(S))$.* (v) *T and V are inverse operations, hence T is an isometry of $C(S)$ onto $C(\Omega(S))$.* (vi) *$\Omega(S)$ is the maximal space with property* (i) *in the sense that if S' is a compact Hausdorff space and if f is a homeomorphism of S with a dense subspace of S', then there exists a continuous mapping F of $\Omega(S)$ onto S' such that $f = F \circ q$. If also every x in $C(S)$ determines a $T'x$ in $C(S')$ such that $T'x(f(s)) = x(s)$ for all s in S, then F is a homeomorphism of $\Omega(S)$ onto S'.*

Proof. (i) follows from boundedness of $q(S)$ and w^*-compactness of bounded, w^*-closed sets; (ii) follows from (i) and the theorem that a continuous image of a compact space in a Hausdorff space is closed. For (iii) V is always a linear mapping of norm 1; it is an isometry because $q(S)$ is dense in $\Omega(S)$. For (iv) observe that each x in $C(S)$ determines a Qx in $C(S)^{**}$ by $Qx(f) = f(x)$ for all f in $C(S)^*$; each Qx is w^*-continuous on $C(S)^*$, so its reduction Tx is continuous on $\Omega(S)$. Both reduction and Q are linear operators of norm 1; since $q(S)$ is dense, V and T are inverse to each other and therefore are isometries onto, so (v) holds.

For (vi) let f be a homeomorphism of S into S' (compact Hausdorff) with $f(S)$ dense in S'. Define V' from $C(S')$ into $C(S)$ by $V'\varphi = \varphi \circ f$; then V' is a linear isometry of $C(S')$ into $C(S)$. Hence V'^* is a w^*-continuous homomorphism of norm 1 of $C(S)^*$ onto $C(S')^*$. If q' is the natural mapping of S' into $C(S')^*$, we need only check that $V^* \circ q = q' \circ f$ to see that $V^*(\Omega(S))$ is a compact dense subset of $\Omega(S')$, which is $q'(S')$ by (ii). Hence $F = q'^{-1} V^*$ satisfies the conditions of (vi). If, finally, $V'C(S')$ is as large as $C(S)$, so that V' is an isometry onto and V'^* is one-to-one, then F is a homeomorphism of $\Omega(S)$ with S'.

Let e be the element of $C(S)$ whose value is one at every element of S.

Corollary 1. *If S is a completely regular space, then $\Omega(S)$ is the set of all those extreme points f of the unit ball U^π of $C(S)^*$ such that $f(e) = 1$.*

Proof. By (v) of the theorem we may replace S by $\Omega(S)$, so we may assume that S is compact. Then (1b) asserts that every element of $q(S)$ is extreme. Trivially every element of $q(S)$ satisfies $f(e) = 1$. By Mil'man's Theorem 1, 3 and (2b) the extreme points of U^π are in the closure of $q(S) \cup -q(S)$. But $q(S)$ is compact, so every extreme point of U^π is either in $q(S)$ or in $-q(S)$. But the points in $-q(S)$ satisfy $f(e)=-1$, so $q(S) = \Omega(S)$ consists precisely of the extreme points for which $f(e)=1$.

From this follows immediately the Banach-Stone theorem; Banach, p. 170, gives the case where S is compact metric; Stone [1] gives the compact Hausdorff case.

Theorem 2 [Banach-Stone]. *If S and S' are two compact Hausdorff spaces, then $C(S)$ and $C(S')$ are linearly isometric if and only if S and S' are homeomorphic.*

Proof. If h is a homeomorphism of S and S', then $Hx' = x' \circ h$ defines a linear isometry H of $C(S)$ with $C(S')$.

If H is a linear isometry of $C(S')$ onto $C(S)$, then H^* is a linear isometry and w^*-homeomorphism of $C(S')^*$ onto $C(S)^*$. Hence H^* carries U^π onto U'^π and the set E of extreme points of U^π onto the corresponding set E' of extreme points of U'^π. Now $E = q(S) \cup -q(S)$ and $E' = q'(S') \cup -q'(S')$; define α on S by $\alpha(s) = (-1)^i$ if and only if $H^* q(s) \in (-1)^i q'(S')$. α is w^*-continuous because

$$\alpha^{-1}((-1)^i) = H^{*-1}((-1)^i q'(S'))$$

is w^*-closed. Define $h(s) = q'^{-1}(\alpha(s) H^* q(s))$; then h is a homeomorphism of S onto S' and $Hx'(s) = \alpha(s) x'(h(s))$ for all x' in $C(S')$ and s in S.

A related result is the Stone-Weierstrass Theorem VI, 2, 1; we give here a useful lemma for a variation of it used in Jerison's theorem in §4.

(4) Call a subspace L of a linear subspace B of $m(S)$ a *normal subspace of B* if for each $x \neq 0$ in B there exists y in L such that $\|y\| = 1$, $y(s) = 1$ if $x(s) \geqq \|x\|/3$, and $y(s) = -1$ if $x(s) \leqq -\|x\|/3$. Suppose that L is a normal subspace of B and that x_0 in B is an element of norm 1; then (a) If y_0 is chosen for x_0 by the normality property, then $\|x_0 - y_0/3\| = 2/3$. (b) If x_i and y_i are already known, let $x_{i+1} = 3(x_i - y_i/3)/2$, so $\|x_{i+1}\| = 1$, and construct y_{i+1} by the defining property of L. For each n, set $z_n = \sum_{i \leqq n} 2^i y_i / 3^{i+1}$. Then $z_n \in L$ and $\|x_0 - z_n\| = (2/3)^{n+1}$. Hence (c) *If L is a normal subspace of a subspace B of $m(S)$, then L is dense in B.*

(5) In any Banach space over the real field, Banach, p. 166, showed that every isometry is a linear isometry followed by a translation. See Chap. VII, § 1 for the proof.

(6) The relations between the real-valued and the complex-valued continuous function spaces discussed in I, 1, (7), I, 2, (6), and I, 4, (18), and II, 1, (10), show that the above theorems are valid for complex-valued functions.

(7) Jerison [1] showed that when the values of the continuous functions are allowed to lie in a general Banach space B instead of in

the real numbers, then it is impossible to conclude that S is homeomorphic with S' whenever $C(S,B)$ is linearly isometric with $C(S',B)$. However, he carried the proof through in the case where B is rotund. (Def. VII, 2, 1.)

(8) The Banach-Stone theorem holds also for completely regular spaces S_1 and S_2 which have at every point a denumerable neighborhood basis, for example, for metric spaces. After the Banach-Stone theorem has been applied to the Stone-Čech compactifications, the original spaces are homeomorphic, because no point of $\Omega(S_i)\setminus S_i$ has a denumerable basis. See also Eilenberg.

§ 4. Characterizations of Continuous Function Spaces among Banach Spaces

Many of the characterizations of continuous function spaces involve directly the partial order relations available in every $C(S)$. In terms of concepts definable in terms of the vector structure and the norm Arens and Kelley gave two characterizations one of which was later improved by Jerison [2] using methods first used by S. B. Myers.

(1) Myers calls a wedge W in B a *T-set* if it is a maximal wedge subject to the restriction that the norm is an additive function in W. Let U be the unit ball in B. If the construction of the Mil'man-Rutman Theorem 1, 3 is applied to the unit ball U^π in (B^*, w^*), then for each f in U^π, $W(f)$, defined to be the set of all F in $(B^*, w^*)^*$ such that $F(f) = \sup F(U^\pi)$, is by Theorem I, 5, 3 the image under Q of $\{x: f(x) = \|x\|\}$. By slight abuse of notation, call this latter set $W(f)$. Then: (a) If $\|f\| < 1$, then $W(f) = \{0\}$. (b) If $\|f\| = 1$, then $W(f)$ may be $\{0\}$ or may not; in any case $W(f)$ is contained in some T-set. (c) If W is a T-set and $K = W \cap \{x: \|x\| = 1\}$, then K is convex and does not meet the interior of U, so, by Eidelheit's Theorem I, 6, 4, there is an f of norm 1 such that $f(x) = 1$ if $x \in K$; hence every T-set is contained in a $W(f)$. (d) Hence, the T-sets are the maximal elements of $\mathfrak{W}$, the set of all $W(f)$, $f \in U^\pi$, so the T-sets of B correspond to the minimal facets of U^π; indeed, the norm in B corresponds to the functional p of the Mil'man-Rutman theory.

(2) If W is a maximal $W(f)$, following Myers, define φ_W on B by $\varphi_W(b) = \inf\{\|b+w\| - \|w\| : w \in W\}$. Now let $\Delta = W - W$, the linear hull of W; then as a partially ordered linear space (I, § 6) Δ is a directed system as is its subsystem W. φ_W has many elementary properties, most of which follow from: (a) For each b in B the function $\|b+x\| - \|x\|$ is a non-increasing function of x in Δ; hence $\varphi_W(b) = \lim_{x \in \Delta}(\|b+x\| - \|x\|)$

$=\lim_{w\in W}(\|b+w\|-\|w\|)$. Then: (b) $\varphi_W(b)\leqq\|b\|$ if $b\in B$. (c) $\varphi_W(b)=\|b\|$ if $b\in W$ and $=-\|b\|$ if $b\in -W$. (d) φ_W is a sublinear functional on B which is linear on Δ. Other properties which are consequences of Lemma 1 below, are: (e) $\varphi_W=\varphi_{W'}$ if and only if $W=W'$. (f) φ_W is continuous. (g) φ_W is linear if and only if $\varphi_W(b)=-\varphi_W(-b)$ for all b in B. (h) φ_W is linear if and only if the function $\|t\|$ has a unique linear extension from W to all of B; that is, if and only if the minimal facet F_W determined by W contains only one point.

Lemma 1. *If F_W is the minimal facet of U^π corresponding to W, then for each b in B*

$$\varphi_W(b)=\sup\{f(b): f\in F_W\}.$$

Proof. Since each f in F_W is of norm one, by (2a)

$$\|b+w\|-\|w\|\geqq f(b+w)-f(w)=f(b);$$

hence $\varphi_W(b)\geqq f(b)$ if $f\in F_W$, so $\varphi_W(b)\geqq\sup\{f(b): f\in F_W\}$.

But φ_W is linear on Δ so, by (2d) and the Hahn-Banach Theorem I, 3, 1, for each b in B there is a linear functional f in B^* dominated by φ_W and therefore agreeing with φ_W on Δ. Moreover if t is any number such that

$$\sup\{-\varphi_W(-y-b)-\varphi_W(y): y\in\Delta\}\leqq t\leqq\inf\{\varphi_W(x+b)-\varphi_W(x): x\in\Delta\}$$

there is such an f with $f(b)=t$. But

$$\begin{aligned}\varphi_W(x+b)-\varphi_W(x)&=\lim_{y\in\Delta}(\|x+b+y\|-\|y\|)-\lim_{y\in\Delta}(\|x+y\|-\|y\|)\\ &=\lim_{y\in\Delta}(\|x+y+b\|-\|x+y\|)=\varphi_W(b).\end{aligned}$$

Hence the restriction on t above simplifies to

$$-\varphi_W(-b)\leqq t\leqq\varphi_W(b).$$

Hence there is an f dominated by φ_W which takes the value $\varphi_W(b)$ at b; this f is also in F_W.

Definition 1. (a) If S is a Hausdorff space and σ is a homeomorphism of S on itself such that $\sigma^2 s=s$ for each s in S, let $C_\sigma(S)$ be the subspace of those functions in $C(S)$ satisfying the conditions $x(\sigma s)=-x(s)$ for all s in S. (b) A linear subspace L of a subspace B of $C(S)$ is called a *separating* [*completely regular*] *subspace* of B if for each point s and set A of points of S for which there exists x in B with $x(s)>\sup|x(A)|$ there is a y in L with $\sup|y(A)|<y(s)$ [and $y(s)=\|y\|$].

(3) In $C(S)$ with S compact Hausdorff, Cor. 3, 1 can be used to show that (a) each maximal wedge W in which norm is additive is

attached to a point s of S; either $W=\{x: x(s)=\|x\|\}$ or

$$W=\{x: x(s)=-\|x\|\};$$

(b) each extreme point of the unit ball in $C(S)^*$ is a φ_W, with W as in (a).

(4) In $C_\sigma(S)$ with S completely regular (a) every point s can be separated from every set A such that s is not in the closure of $A \cup \sigma A$. (b) If also S is compact Hausdorff, the extreme points φ of U^π and the maximal wedges in $C_\sigma(S)$ where norm is additive match one-to-one with the points s of S where $s \neq \sigma s$, under the relations $\varphi_s(x)=x(s)$ for all x in $C_\sigma(S)$ and $W_{\varphi(s)}=\{x: x(s)=\|x\|\}$. (Use (3a) and shift, if necessary, from s to σs.)

The next theorem combines results of Myers and of Jerison [2].

Theorem 1. (a) *Every Banach space is isometric to a separating subspace of a space $C_\sigma(S)$ with S completely regular. One suitable space for S is the set of all extreme points of the minimal facets of the unit ball U^π of B^* in its relative w^*-topology.*

(b) *In order that B be isometric to a completely regular subspace of some $C_\sigma(S)$ with S completely regular it is necessary and sufficient that* (i) *there exists in the set of extreme points of U^π which are also one-point minimal facets of U^π a symmetric subset E so large that U^π is the w^*-closed convex hull of E, or* (ii) *there exists a set E of linear functionals φ_W so large that U^π is the w^*-closed convex hull of E.*

(c) *In order that B be isometric to a completely regular subspace of some $C_\sigma(S)$ with S compact Hausdorff it is necessary and sufficient that* (iii) *every φ_W is linear, and* (iv) *the set Φ of all φ_W or $\Phi'=\Phi\cup\{0\}$ is compact in the relative w^*-topology. In this case, if S' is the space obtained by identifying all points of $S_1=\{s: \sigma s=s\}$, then S' is homeomorphic to Φ if $S_1=\emptyset$, to Φ' if $S_1 \neq \emptyset$.* (This compounds Myers's and Jerison's versions of the Banach-Stone theorem.)

Proof of (a) Since the space S defined in (a) is completely regular and since the Mil'man-Rutman Theorem 1.2 asserts that the natural mapping T, defined by $Tx(s)=s(x)$ for all s in S, is an isometry, it remains only to show that $T(B)$ is separating in $C_\sigma(S)$. Let A be any subset of S and let s be a point not in $w^*(A\cup -A)$. By "(iv) implies (ii)" of the K^2-M^3-R theorem of Section 1, there is an x in B such that

$$Tx(s) > \sup Tx(A\cup -A) = \sup|Tx(A)|;$$

that is, $T(B)$ is separating in $C_\sigma(S)$.

Proof of (b). A set E satisfying either (i) or (ii) also satisfies the other condition as well, by Lemma 1. Define $Tx(\varphi)=\varphi(x)$ for every φ in E, as before; to prove $T(B)$ completely regular in $C_\sigma(E)$ let A be a subset of E and s an element of E such that $s \notin w^*(A \cup -A) = A'$. We desire a w in B such that $\|w\| = Tw(s) > \sup|Tw(A)| = \sup Tw(A')$. If no such w exists, then for every w in $W(s)$, $\sup Tw(A') = s(w)$. Let $F_w = \{\varphi: w \in W(\varphi)\}$; the property assumed for w and compactness of A' yield that $F_w \cap A' \neq \emptyset$. If this holds for every w in $W(s)$, which is non-empty, then the $F_w \cap A'$ have the finite intersection property; for if $w = \sum_{i \leq n} w_i$ and $w_i \in W(s)$, then $w \in W(s)$ and $F_w = \bigcap_{i \leq n} F_{w_i}$, so $\bigcap_{i \leq n} (F_{w_i} \cap A') = F_w \cap A' \neq \emptyset$. By compactness $F_{W(s)} \cap A' = \bigcap_{w \in W(s)} (F_w \cap A') \neq \emptyset$. But this contradicts the fact that $F_W = \{s\}$ and $s \notin A'$. Hence $T(B)$ is a completely regular subspace of $C_\sigma(E)$ if E is chosen as in (i) or (ii).

Conversely, if T is a linear isometry of B into a completely regular subspace of some $C_\sigma(S)$, with S completely regular, let q be the natural mapping of S into $C_\sigma(S)^*$; $qs(f) = f(s)$ for all f in $C(S)$. Then $T^* qs \in B^*$ and, since $T(B)$ is completely regular in $C_\sigma(S)$, as in the proof of (4b) $T^* qs$ is of norm one if $s \neq \sigma s$. Then

$$W(T^* qs) = \{x: \|x\| = T^* qs(x)\} = \{x: \|Tx\| = Tx(s)\}$$

is a set in B where norm is additive.

We need next to prove:

(v) If $\sigma s \neq s$, then $W(T^* qs)$ is a maximal wedge in B in which norm is additive and $\varphi_W = T^* qs$.

To prove this observe that every w for which $\|w\| = Tw(s)$ is in W by definition. If an x in B satisfies $1 = \|x\| > Tx(s)$, then by complete regularity of $T(B)$ in $C_\sigma(S)$ there exist $\delta > 0$ and w in W such that if $|x(s')| \geq 1 - \delta$, then $|w(s')| \leq 1 - \delta$ hence $\|x + w\| < \|x\| + \|w\|$ and norm is not additive on the enlarged wedge containing W and x. $T^* qs \in F_W$, so by Lemma 1, $T^* qs(x) \leq \varphi_W(x)$ for all x in B.

To prove the opposite inequality, take $\varepsilon > 0$ and a neighborhood N of s small that $|Tx(s) - Tx(s')| < \varepsilon$ if $s' \in N$. Then there is w in W such that $1 = \|w\| = Tw(s) > \sup\{Tw(s''): s'' \notin N \cup -N\} = 1 - \delta$. If $k\delta > \|x\|$, then for each $s'' \notin N \cup -N$

$$|T(kw - x)(s'')| \leq k - k\delta + \|x\| \leq k = \|kw\|,$$

while if $s' \in N$,

$$|T(kw + x)(s')| \leq Tkw(s) + Tx(s) + \varepsilon \leq \|kw\| + Tx(s) + \varepsilon$$

so $\|kw + x\| - \|kw\| \leq Tx(s) + \varepsilon$. This holds for every $\varepsilon > 0$; by (2a), $\varphi_W(x) = Tx(s) = T^* qs(x)$.

Property (v) shows that if $E = \{T^* qs: s \neq \sigma s\}$, then E consists of one-point minimal facets of U^π. But T is an isometry into, so $\|x\| = \|Tx\| = \sup\{|Tx(s)|: s \in S\} = \sup\{|T^* qs(x)|: s \in S\} = \sup\{|\varphi(x)|: \varphi \in E\}$ for $T^* qs(x) = 0$ if $s = \sigma s$.

Proof of (c). Suppose that $T(B)$ is completely regular in $C_\sigma(S)$, S compact Hausdorff; let U and U_1 be the unit balls in B and $C_\sigma(S)$, respectively. Then for every extremal set A in U^π, $T^{*-1}(A)\cap U_1^\pi$ is non-empty, by the Hahn-Banach theorem, and is extremal in U_1^π. By the Kreĭn-Mil'man theorem every extreme point of U^π is the image under T^* of an extreme point of U_1^π, then by (4b) the set of extreme points of U^π is contained in $T^*q(S)$. But by part (b) of this theorem every element T^*qs, $s \neq \sigma s$, is a one-point minimal facet of U^π and is a φ_W, W a maximal wedge in B. Hence $\Phi=\{T^*qs: s\neq\sigma s\}$ and $\{0\}=\{T^*qs: s=\sigma s\}$.

If (iii) and (iv) hold, (b) asserts that $T(B)$ is completely regular in $C_\sigma(\Phi)$, where $\sigma\varphi=-\varphi$ for all φ in Φ, so $T(B)$ is still completely regular in $C_\sigma(\Phi')$, and Φ' is compact.

(5) Examples for Theorem 1. (a) $c_0(\omega)$ is a completely regular subspace of $m(\omega)$, but is not completely regular in $C(S)$ for any compactification S of the discrete space ω. If Φ is chosen in $c_0(\omega)^*$ as in (c) of the theorem, $Tc_0(\omega)$ is a completely regular subspace of $C_\sigma(\Phi)$. (b) If $B=l^1(\omega)$, then Φ, defined as in (c) of the theorem, is homeomorphic to the Cantor discontinuum (see (6)) and is, therefore, compact. $Tl^1(\omega)$ is a completely regular subspace of $C_\sigma(\Phi)$.

As a step in the characterization of $C(X)$ among Banach spaces, let us give Jerison's characterization of $C_\sigma(X)$.

Definition 2. A Banach space B has *property* $\mathscr{A}$ (Arens-Kelley) if and only if for each family Γ of maximal convex subsets of the surface of the unit ball U such that $\bigcap_{C\in\Gamma} C=\emptyset$ there exists nets $(C_n, n\in\Delta)$ and $(C_n', n\in\Delta)$ of elements of Γ such that for each b in U

$$\lim_{n\in\Delta}(d(b,C_n)+d(b,C_n'))=2.$$

B is said to have *property* $\mathscr{A}'$ (Jerison [2]) if and only if for each family Ω of maximal wedges $W(f)$ such that $\bigcap_{W\in\Omega} W=\{0\}$ there exist nets $(W_n, n\in\Delta)$ and $(W_n', n\in\Delta)$ in Ω such that for each b in B

$$\lim_{n\in\Delta}(\varphi_{W_n}(b)+\varphi_{W_n'},(b))=0.$$

Theorem 2 (Jerison [2]). *If B is a Banach space, let S_1 be the set of all those extreme points of U^π, the unit ball, of B^*, which are in minimal facets of U^π. The following conditions on B are equivalent:*

(i) *There is a compact Hausdorff space X and an involutory homeomorphism σ of X on itself such that B is isometric to $C_\sigma(X)$.*

(ii) *B has property $\mathscr{A}$.*

(iii) *B has property $\mathscr{A}'$.*

(iv) *For each subset A of S_1 such that $w^*(A)\cap w^*(-A)=\emptyset$ there is a non-zero x in $\bigcap_{s\in A} W(s)$.*

(v) *For each set A as in* (iv) *there is $x \neq 0$ in B such that $f(x) = \|x\|$ for all f in $w^*(A)$.*

(vi) *$T(B)$ is a normal subspace of $C_\sigma(w^*(S_1))$.*

(vii) *$T(B)$ is all of $C_\sigma(w^*(S_1))$.*

Proof. (ii)$\Rightarrow$(iii). Take b in U, W a maximal $W(s)$, and C the set of points of norm 1 in W, and use the inequalities

$$1 - d(b, C) \leqq \varphi_W(b) \leqq d(-b, C) - 1 .$$

(iii)$\Rightarrow$(iv). Begin with the fact that $\{W, -W\}$ is a family Ω with intersection $= \{0\}$. $\mathscr{A}'$ implies that $\varphi_W(b) + \varphi_{-W}(b) = 0$ if $\|b\| \leqq 1$, but $\varphi_{-W}(b) = \varphi_W(-b)$, so (2h) asserts that φ_W is linear. Then every minimal facet of U^π has a single point in it. Theorem 1 (b) then asserts that B is isometric to a completely regular subspace of $C_\sigma(S_1)$.

Now take A such that $w^*(A) \cap w^*(-A) = \emptyset$. If $\bigcap_{s \in A} W(s) \neq \{0\}$, take x of norm 1 in that intersection; then $f(x) = 1$ for all f in $w^*(A)$. But in the presence of $\mathscr{A}'$ it can not happen that the intersection is $\{0\}$; for if it were, there would exist (s_n), (s'_n) in A such that $\lim_{n \in \Delta}(s_n(b) + s'_n(b)) = 0$ for all b in U; then if s is the w^*-limit of some subnet of (s_n), $-s$ is the w^*-limit of the corresponding subnet of (s'_n) so s is in $w^*(A) \cap w^*(-A)$, contrary to the choice of A.

(iv)$\Rightarrow$(v). Use continuity of each Tx on $Y = w^*(S_1)$.

(v)$\Rightarrow$(vi). If $\xi \in C_\sigma(Y)$, the set $A = \{s : \xi(s) \geqq \|\xi\|/3\}$ satisfies the hypothesis of (v).

(vi)$\Rightarrow$(vii). Use § 3, (4).

(vii)$\Rightarrow$(i). Take $X = w^*(S_1)$, $\sigma = -$.

(i)$\Rightarrow$(ii). If (i) holds, each maximal wedge W in B is determined by an element t of X; $W = W_t = \{x : x(t) = \|x\|\}$. Then if $C_t =$ the set of points of norm one in W_t, and if $b \in U$, it follows that $d(b, C_t) = 1 - b(t)$. [To get a nearest b' in C_t add $e(1 - b(t))$ to b and trim the result off at 1.] Hence if A is a set in X such that $\bigcap_{t \in A} C_t = \emptyset$, A and $\sigma(A)$ could not have disjoint closures, for if they had there would, by normality, be g in $C(X)$ with $\|g\| = 1$, $g(t) = 1$ if $t \in A$, $g(t) = -1$ if $t \in \sigma(A)$; then $f = (g - g \circ \sigma)/2$ would be in $C_\sigma(X)$ and $f(t) = \|f\|$ if $t \in A$, contrary to the supposition that $\bigcap_{t \in A} C_t = \emptyset$. Hence there exists nets (t_n) and (t'_n) in A such that $w^*\text{-}\lim_{n \in \Delta} t_n = w^*\text{-}\lim_{n \in \Delta} -t'_n$; the corresponding C's fit the conclusion of $\mathscr{A}$.

Corollary 1 [Arens-Kelley-Jerison]. *B is isometric to a $C(X')$ with X' compact Hausdorff, if and only if* (i) *B has property $\mathscr{A}$ (or any other of the properties of the theorem), and* (ii) *the unit ball U of B has at least one extreme point.*

Proof. Define $X = X' \cup X''$, where X'' is a homeomorph of X' and both sets are closed in X; then $C(X')$ is isometric to $C_\sigma(X)$; by the theorem $C(X')$ has property $\mathscr{A}$. Also any function on X' of constant absolute value 1 is an extreme point of U.

If, on the other hand, B has $\mathscr{A}$ and U has an extreme point e, B may be regarded as a $C_\sigma(X)$, with X the w^*-closure of the set of extreme points of U^π. For each extreme point s of U^π either $s(e) = 1 = \|e\|$ or $s(e) = -1 = -\|e\|$. The two sets X' and X'' thus defined have disjoint w^*-closures and $\sigma(X') = -X' = X''$. Hence $C_\sigma(X)$ is isometric to $C(X')$.

Corollary 2 (Jerison [2]). *B is isometric to the space $C_0(X)$ of all continuous functions vanishing at infinity on a locally compact Hausdorff space X if and only if B has property $\mathscr{A}$ and the set Φ of extreme points of the unit ball in B^* is the union of two disjoint closed antipodal subsets Φ_1 and $-\Phi_1$.*

If U in B has an extreme point, this returns to Cor. 1 and Φ_1 is compact; if U has no extreme point, use Th. 2 and Th. 1 (c).

(6) (a) There is a simple characterization of the family of all linear subspaces of $C(S)$ spaces, S compact Hausdorff: *Every normed space N is linearly isometric to a linear subspace of $C(S)$, where S is the w^*-topologized unit ball of N^*.* (b) By II, 5, (4c) if N is separable, S is compact metrizable. Hence it is a continuous image of the Cantor discontinuum 2^ω [Kelley, p. 166]. Hence every separable normed space is linearly isometric to a linear subspace of $C(2^\omega)$. (c) 2^ω is homeomorphic to a closed subset of $[0,1]$. Extending each continuous function on this subset linearly across each open interval of the complement gives a linear isometry of $C(2^\omega)$ into $C([0,1])$; hence: *Every separable normed linear space is linearly isometric to a linear subspace of* $C([0,1])$ [Banach, p. 163].

(7) The extension device used in (6c) is greatly generalized in a paper of Kakutani [1] on simultaneous extension of all continuous functions defined on a closed subset of a separable metric space to continuous functions defined on the whole space. E. Michael improved this result and gave examples (see also Day [5]) to show the restrictions on the possibility of such extensions over more general topological spaces.

(8) (a) If S is compact Hausdorff and σ is an involutory homeomorphism of S and if $s \neq \sigma s$, then for any open G about s such that $\sigma G \cap G = \emptyset$, take x in $C(S)$ such that $\|x\| = x(s) = 1$ and $x(s') = 0$ if $s' \notin G$. If $y = x - x \circ \sigma$, then $y \in C_\sigma(S)$ and $y \in W(\varphi_s)$, where φ_s is the evaluation functional at s: $\varphi_s(x) = x(s)$ for all x in $C_\sigma(S)$. (b) Norm in $C_\sigma(S)$ is not additive unless the functions added have a common maximum point. (c) The maximal wedges of $C_\sigma(S)$ are all of the form $W(\varphi_s)$. Part (a) shows that if $s \neq \sigma s$, and $s' \neq s$, there is x in $W(\varphi_s)$

such that $x \notin W(\varphi_{s'})$. Hence the extreme point φ_s of U^π is the only element in its minimal facet of U^π.

This can now be applied to a special type of Banach space studied by Akilov [1,2], Goodner, Nachbin [1], and Kelley [2].

II, 1, (4a) implies that every 1-dimensional subspace B of every Banach space B' is the range of a projection of B' on B with norm one. We shall characterize spaces with this "onto" projection property and give several equivalent properties involving extension of operators without increase of norm. All of these results are analogous to the monotone extension theory of VI, § 3.

Definition 3. For each $\lambda \geq 1$, let $\mathfrak{P}[\mathfrak{P}_\lambda]$ be the family of those Banach spaces B such that for each Banach space $B' \supseteq B$ there is a continuous projection; that is, a linear idempotent mapping P, carrying B' onto B [with $\|P\| \leq \lambda$].

(9) (a) Phillips [1] observed that the Hahn-Banach theorem applied coordinatewise shows that *for every index set* $S, m(S) \in \mathfrak{P}_1$. (b) If $B \in \mathfrak{P}_\lambda$, the definition $F = f \circ P$ shows that B also satisfies: If $B \subseteqq Y$, a normed linear space, then every continuous linear operator f from B into a normed X has a linear extension F with $\|F\| \leqq \lambda \|f\|$. Setting $X = B$, $f =$ identity, gives the converse. (c) Imbedding B into an $m(S)$ shows, with (a), that when $B \in \mathfrak{P}_\lambda$ then for every normed linear space $X \subseteqq$ a normed linear Y and every continuous linear f from X into B there is a continuous linear extension F from Y into B such that $\|F\| \leqq \lambda \|f\|$. The converse also holds. (d) Another condition equivalent to $B \in \mathfrak{P}_\lambda$ is: If f is a linear isometry of B into a normed linear Y, then there is a g from Y onto B such that $\|g\| \leqq \lambda$ and $g \circ f$ is the identity operator in B. [For if (c) holds, f^{-1} has such an extension g; if f is the identity and g exists with this property, then $P = f \circ g$ is a projection of Y on B which shows that $B \in \mathfrak{P}_\lambda$.] (e) $\mathfrak{P} = \bigcup_{\lambda \geq 1} \mathfrak{P}_\lambda$. [For λ use the norm of a projection of $m(U^\pi)$ onto B.] See Dean or Sobczyk [3].

These results can be found in Goodner and, in part, in Akilov [1,2]; Nachbin [1] and Goodner characterized those normed spaces in $\mathfrak{P}_1$ which have an extreme point on the unit ball. J.L. Kelley showed that this last assumption is extraneous; we give here a variant of Kelley's proof [2] that (a) implies (b).

Definition 4. A space B has the binary intersection property if for each family $\{V_z, z \in Z\}$ of closed balls in B for which each pair intersects there is a point common to all the V_z.

Theorem 3. *The following conditions on a Banach space B are equivalent:*

(a) $B \in \mathfrak{P}_1$.

(b) *B is isometric isomorphic to a space $C(S)$ where S is compact, Hausdorff, and extremally disconnected*[1].

(c) *B has the binary intersection property.*

Proof. (a) implies (b). Suppose that $B \in \mathfrak{P}_1$ and begin, as in earlier theorems of this section, by imbedding B in $C_\sigma(S_1)$, where S_1 is the closure of the set of extreme points of the unit ball U^π of B^* in the relative w^*-topology, and $\sigma s = -s$, by the mapping $Tx(s) = s(x)$ for all x in B. Then S_1 is compact Hausdorff, so the extreme points of the unit ball V^π in $C_\sigma(S_1)^*$ are the evaluation functionals φ_s, one for each point of S_1, with the exception of 0 if that belongs to S_1. There is a natural map $q: s \to \varphi_s$ of S_1 into $C_\sigma(S_1)^*$ defined by $q s(Tx) = \varphi_s(Tx) = Tx(s) = s(x)$ for all x, s.

Now consider the set $K = T^{*-1}(s) \cap V^\pi$; K is extremal in V^π if s is an extreme point of U^π; hence each extreme point of K is extreme in V^π and is, therefore, an evaluation functional $\varphi_{s'}$. But $T^* \varphi_{s'} = s$ or $s(x) = T^* \varphi_{s'}(x) = \varphi_{s'}(Tx) = s'(x)$ for all x in B, so $s' = s$; by the Kreĭn-Mil'man theorem, when s is extreme in U^π, the only point of $T^{*-1}(s) \cap V^\pi$ is φ_s.

$B \in \mathfrak{P}_1$ so (9e) gives a linear function g from $C_\sigma(S_1)$ onto B such that $g \circ T = \mathfrak{i}$, the identity operator in B. Hence $T^* \circ g^* = \mathfrak{i}^*$. If s is extreme in U^π it follows that $g^*(s) \in K$ or $g^*(s) = q(s)$ for each extreme point s in U^π; since g^* and q are both continuous, $g^*(s) = q(s)$ for all s in S_1.

If A is a closed subset of S_1 such that $A \cap -A = \emptyset$, then in $C_\sigma(S_1)^*$, $q(A) \cap \sigma q(A) = \emptyset$, so, by (v) of Theorem 2, there is an x in $C_\sigma(S_1)$ such that $x(s) = 1$ for all s in A, so $x(s) = -1$ for all s in $-A$. Then

$$1 = x(s) = q s(x) = g^* s(x) = s(g x)$$

for all s in A. This proves that $T(B)$ has the property of (v) of Theorem 2; that is, $T(B)$ is all of $C_\sigma(S_1)$.

Next let G be a maximal open subset of S_1 such that $G \cap -G = \emptyset$. In $m(S_1)$ consider the normed linear subspace B' generated by $B = C_\sigma(S_1)$ and χ, where $\chi(s) = 1$ if $s \in G$, $\chi(s) = -1$ if $s \in -G$, and $\chi(s) = 0$ elsewhere. Let P be a projection of norm one of B' on B and let $x = P\chi$. Clearly $\|x\| \leqq \|\chi\| = 1$. If $s \in G$, there is a function y in B such that $\|y\| = 2 = y(s)$, $y(s') = 0$ except for s' in $G \cup -G$, and $y(s') \geqq 0$ if $s' \in G$. Then $\|\chi - y\| = 1$ so

$$1 \geqq \|P\chi - Py\| = \|x - y\| \geqq x(s) - y(s) = x(s) - 2 \geqq -1 .$$

Hence $x(s) = 1$ if $s \in G$, $x(s) = -1$ if $s \in -G$ and $x \in C_\sigma(S_1)$. Since $G \cup -G$ is dense in S_1, $x(s) = 1$ or -1 for all s in S_1; therefore S, the

[1] See also VI, §3; S is called *extremally disconnected* if and only if the closure of every open sets is open.

closure of G, is also open in S_1 and $S \cup -S = S_1$. By Corollary 1, B is isometric with $C(S)$.

Repetition of the above argument with any open set G' in S and the interior G'' of its complement in S shows that the closure of every open set in S is open.

This completes the proof that if $B \in \mathfrak{P}_1$, then B is isomorphic to a $C(S)$ with S compact, Hausdorff, and extremally disconnected. A direct proof of the converse will be given in VI, 3, (2) after the discussion of partially ordered normed spaces. It depends on use of the Hahn-Banach theorem for functions with values in $C(S)$.

(b) implies (c). Every $C(S)$ is a lattice under pointwise operations, and VI, 3, (2a) says that S is extremally disconnected if and only if $C(S)$ is a boundedly complete lattice. If V_z, z in Z, is a family of closed balls in $C(S)$ which intersect in pairs, then each V_z is an order interval $\{x: a_z \leq x \leq b_z\}$, and for each finite set $\zeta \subset Z$, $V_\zeta = \bigcap_{z \in \zeta} V_z$ is the order interval from $a_\zeta = \sup_{z \in \zeta} a_z$ to $b_\zeta = \inf_{z \in \zeta} b_z$. Then $\sup_{\zeta \subset Z} a_\zeta \leq \inf_{\zeta \subset Z} b_\zeta$, so $\bigcap_{z \in Z} V_z$ is the interval with these ends, and $C(S)$ has the binary intersection property.

(c) implies (a). In the hypotheses of (9b) suppose that f is defined, linear, and of norm one into B from some subspace Z of a normed space Y. For a fixed y in $Y \setminus Z$ adapt the Hahn-Banach proof from Th. I, 3, 1 by defining for each z in Z, V_z to be the closed ball about $f(z)$ of radius $\|z-y\|$. Pairs of these V_z intersect, so there is some b in B for which $\|f(z)-b\| \geq \|z-y\|$ for all z in Z. The definition $g(z+ty) = f(z)+tb$ yields a linear extension of f of norm one defined on a subspace of Y larger than Z; this is the basic induction step so the rest goes as in Th. I, 3, 1.

(10) No corresponding characterization for spaces in $\mathfrak{P}_\lambda$, $\lambda > 1$, is known.

(11) Akilov [2] proved that if the unit ball in B is smooth (Definition VII, 2, 1), in particular, if B is L^p or $l^p(\omega)$, with $p > 1$, then B is not in $\mathfrak{P}_1$.

(12) If B is reflexive and is in $\mathfrak{P}_\lambda$, then B is finite-dimensional. [For B is isometrically embeddable in $m(U^\pi)$. Any projection P of $m(U^\pi)$ on B is *wcc* by Theorem III, 4, 1. By the Dunford-Pettis theorem VI, 4, 5, $P = P^2$ is *cc*; hence its range B is finite-dimensional by Theorem II, 5, 1.]

(13) (a) Phillips, by his Lemma II, 2, 1, showed that there is no continuous linear projection of $m(\omega)$ onto $c_0(\omega)$. (b) Sobczyk [1] showed that $c_0(\omega)$ has a weak form of property $\mathfrak{P}_2$; every *separable normed* space containing $c_0(\omega)$ can be projected on it with norm $\leqq 2$.

A simple proof is now available in Veech.

Chapter VI. Norm and Order

§ 1. Vector Lattices and Normed Lattices

In this section we discuss elementary properties of vector lattices with the ultimate goal of characterizing continuous function spaces and their closed vector sublattices.

Definition 1. Let V be a partially ordered linear space with a positive wedge W. (Def. I, 6, 2.) (a) If A is a subset of V, then an element b of V is an *upper* [*lower*] *bound for* A if $b \geqq a$ [$b \leqq a$] for all a in A. (b) b is *a least upper* [*greatest lower*] *bound for* A if it is such an upper [lower] bound for A that whenever b' is an upper [lower] bound for A then $b' \geqq b$ [$b' \leqq b$]. (c) V is a *vector lattice* if the positive wedge in V is a cone and if each set $\{x, x'\}$ of two elements has a least upper bound, which will be denoted by $x \vee x'$. (d) A vector lattice is *boundedly* [σ-] *complete* if each [countable] set A which has an upper bound has a least upper bound, $\sup A$. (e) A *vector sublattice* V_0 of a vector lattice V is a linear subspace of V such that if $x, x' \in V_0$, then $x \vee x' \in V_0$.

(1) Let V be a vector lattice and let $x, y, z \in V$, $\lambda, \mu \in R$. Then: (a) $\{x, y\}$ has a greatest lower bound $-((-x) \vee (-y))$; denote it by $x \wedge y$. (b) $(x \vee y) + z = (x+z) \vee (y+z)$. (c) $\lambda(x \vee y) = \lambda x \vee \lambda y$ if $\lambda \geqq 0$. (d) $(x \wedge y) + z = (x+z) \wedge (y+z)$. (e) $\lambda(x \wedge y) = \lambda x \wedge \lambda y$ if $\lambda \geqq 0$. (f) $(x \vee y) + (x \wedge y) = x + y$.

(2) Let V be a vector lattice with positive cone K; for each x in V define $x^+ = x \vee 0$, and $x^- = (-x)^+ = -(x \wedge 0)$. Then (a) $x^+ - x^- = x$. (b) If $x = y - z$ with y, z, in K, then $y - x^+ = z - x^- \in K$. (c) The function p defined from V into K by $p(x) = x^+$ for all x is a sublinear function. (d) $x \vee y = (x-y)^+ + y$. (e) If x and y are non-zero and $x \wedge y = 0$, then $\lambda x + \mu y \in K$ if and only if $\lambda \geqq 0$ and $\mu \geqq 0$; hence (f) if $x \wedge y = 0$, then $(\lambda x + \mu y)^+ = \lambda^+ x + \mu^+ y$. (g) An f in $V^\#$ is a lattice homomorphism (that is, preserves both $\vee$ and $\wedge$) if and only if $f(x \vee y) = f(x) \vee f(y)$ whenever $x \wedge y = 0$.

Definition 2. If $E \subseteqq V$, an OLS, then $E^+ = \{f : f \in V^\#$ and $f(x) \geqq 0$ for all x in $E\}$.

(Note that if E is a wedge, then, by I, 6, (11), $E^+ = E^\pi$).

(3) Let V be a vector lattice with positive cone K; let x, y be linearly independent points of V for which $x \wedge y = 0$, let $f \in K^+$, and define p on V by $p(z) = f(z^+)$ for all z in V; then: (a) p is non-negative and sublinear, and $p(z) = 0$ if $z \leqq 0$. (b) If f_0 is defined in the plane of x and y by $f_0(\lambda x + \mu y) = \lambda f(x)$, then f_0 is dominated by p. (c) Hence f_0 has a linear extension f_1 defined on V and dominated by p. (d) Hence $f_1 \in K^+$ and $f_2 = f - f_1 \in K^+$, $f_1(x) = f(x)$ and $f_2(y) = f(y)$.

(4) If V is a vector lattice, call an element f of K^+ *indecomposable* if $f = f_1 + f_2$ and f_i in K^+ imply that $f_i = \lambda_i f$, with $0 \leqq \lambda_i \leqq 1$. (2g) and (3) show that *if f is not a lattice homomorphism, then f is decomposable.* To prove *the converse* let f be a lattice homomorphism and let $f = f_1 + f_2$, with the f_i in K. (a) For each x in V, either $f(x^+) = 0$ or $f(x^-) = 0$. (b) If $f(x) = 0$, then $f(x^+) = f(x^-) = 0$; hence $f_i(x^+) = f_i(x^-) = 0$, so (c) $f_i(x) = 0$ if $f(x) = 0$. (d) By I, 2, (3d), $f_i = \lambda_i f$; restrictions on the λ_i are a consequence of the relative sizes of f and f_i.

We follow the terminology of Kreĭn and Rutman in the next definitions, the word "minihedral" is suggested by the finite-dimensional situation; see (7e).

Definition 3. A cone K in a partially ordered vector space V is called (a) *minihedral* if each two-element set in K has a least upper bound; (b) *fully* [σ-] *minihedral* if each [countable] set in K which is bounded above has a least upper bound; (c) *reproducing* if $K - K = V$.

(5) (a) If K is a minihedral cone, if $x_1, x_2 \in K$, and if $0 \leqq y \leqq x_1 + x_2$, then setting $y_1 = x_1 \wedge y$, $y_2 = y - y_1$ gives $0 \leqq y_i \leqq x_i$ for $i = 1$ and 2. [Use (1f) to show that $x_1 + y_2 \leqq x_1 + x_2$.] (b) From (a) it follows in a vector lattice when $0 \leqq x_i$ and $0 \leqq c$ that $(x_1 \wedge c) + (x_2 \wedge c) \geqq (x_1 + x_2) \wedge c$. (To prove this, use the right hand side here in place of y in (a).)

Definition 4. If V is a partially ordered linear space, let V' be the set of elements of $V^\#$ which are bounded on every order interval, $\{x : a \leqq x \leqq b\}$ of V.

(6) If K is a minihedral cone then $K^+ \subseteq V'$, and if $f \in V'$, then there is an f^+ in K^+ such that $f^+ = f \vee 0$; that is, if $V^\#$ is ordered by the wedge K^+, then $f^+ \geqq f$, $f^+ \geqq 0$, and if g in $V^\#$ is an upper bound for f and 0, then $f^+ \leqq g$. This f^+ is defined stepwise as follows: In K set $f^+(x) = \sup \{f(y) : 0 \leqq y \leqq x\}$; then: (a) $f^+(\lambda x) = \lambda f^+(x)$ if $\lambda \geqq 0$ and $x \in K$. (b) By (5), $f^+(x+y) = f^+(x) + f^+(y)$ if x and $y \in K$. (c) Define $f^+(x-y) = f^+(x) - f^+(y)$ if $x, y \in K$; then: (b) implies that if $z = x - y = x' - y'$, then $f^+(x+y') = f^+(x'+y)$, so f^+ is defined in $K - K$ and is linear there. (d) Any linear extension of f^+ to all of V has the required properties. Hence: (e) If V is a vector lattice, then V' is a vector lattice.

(f) For each x in K and f, g in V',

$$(f \vee g)(x) = \sup\{g(y) + f(z) : y \text{ and } z \in K \text{ and } y + z = x\}.$$

(7) (a) Every $C(S)$ is a vector lattice. (b) $C([0,1])$ is not a boundedly σ-complete vector lattice. (c) If S has the discrete topology, so that $C(S) = m(S)$, then $C(S)$ is a boundedly complete vector lattice. (d) A partially ordered linear space V is a vector lattice if and only if its set K of positive elements is a minihedral, reproducing cone in V. (e) In an n-dimensional vector space a minihedral, reproducing cone is a set $\{x : f_i(x) \geqq 0,\ i = 1, \ldots, n\}$, where the f_i are linearly independent; that is, K has exactly n faces and has non-empty interior.

Definition 5. A *normed* [*Banach*] *lattice* is a normed linear [Banach] space which is also a vector lattice in which $\vee$ and $\wedge$ are continuous functions of both their variables. An (AB)-*lattice* is a normed linear space and a vector lattice in which order and norm are related by the following conditions:

(A) If $x \wedge y = 0$, then $\|x + y\| = \|x - y\|$.

(B) If $0 \leqq x \leqq y$, then $0 \leqq \|x\| \leqq \|y\|$.

(8) (a) Every $C(S)$ is an (AB)-lattice. The cone K in $C(S)$ has as its interior points all x for which $\inf x(S) > 0$. (b) If S is locally compact, Hausdorff, and if $C_0(S)$ is the space of functions in $C(S)$ which vanish at infinity, then $C_0(S)$ is also an (AB)-lattice but its positive cone has no core point. $c_0(S)$ is a simple example. (c) All the spaces $l^p(S)$, with $p \geqq 1$, and with K containing the non-negative elements of $l^p(S)$, are (AB)-lattices.

(9) (a) In a normed lattice V, (A) is equivalent to

(A′) for all x in V, $\|x^+ + x^-\| = \|x\|$,

and (B) is equivalent to

(B′) if x and $y \in K$, then $\|x \vee y\| \geqq \|x\| \vee \|y\|$.

(b) (B) implies that order intervals are norm-bounded, so $V^* \subseteqq V'$. (c) If the cone K in a normed lattice satisfying (A) is complete under convergence of monotone Cauchy sequences, then $V' \subseteqq V^*$. [Kaplansky, conversation about 1949. If it could happen that for f in K^+, $f(U)$ is unbounded, then $f(U \cap K)$ is unbounded by (a) so there exist x_n in $U \cap K$ such that for all $n, f(x_n) > 4^n$. Then $x = \sum_n x_n/2^n$ is in $K \cap U$ and $f(x) \geqq f(x_n/2^n) > 2^n$ for all n!] (d) Continuity of $\vee$ implies that the positive cone in a normed lattice is closed.

Lemma 1. *An* (AB)-*lattice is a normed lattice.*

By (2d), for x, y in V, $x \vee y = (x - y)^+ + y$; the vector operations in V are continuous in both their variables, so continuity of $\vee$ is equivalent to continuity of $(\ )^+$. But by sublinearity $v=x^+-y^+\leqq(x-y)^+$ and $-v=y^+-x^+\leqq(y-x)^+=(x-y)^-$. Hence

$$\|x^+-y^+\|=\|v\|=\|v^++v^-\|\leqq\|(x-y)^++(x-y)^-\|=\|x-y\|.$$

(10) Let V be an (AB)-lattice with positive cone K and unit ball U; then: (a) If $f\in K^+$, then $\|f\|=\sup\{f(x): x\in K\cap U\}$. (b) Hence if $f\in V^*$, $\|f^+\|\leqq\|f\|$ and $f^+\in V^*$. (c) Hence V^* is a vector lattice and (d) V^* satisfies condition (B).

Lemma 2. *If V is an* (AB)*-lattice, then V^* is an* (AB)*-lattice in which every set with an upper bound has a least upper bound.*

Proof. (10) asserts that V^* is a Banach lattice with property (B). To prove that V^* has property (A), use (10a) and (6) to get

$$\begin{aligned}\|f^++f^-\| &= \sup\{f^+(x)+f^-(x): x\in U\}\\ &= \sup\{f^+(x)+f^-(x): x\in U\cap K\}\\ &= \sup\{\sup\{f(y): 0\leqq y\leqq x\}\\ &\quad +\sup\{-f(z): 0\leqq z\leqq x\}: x\in K\cap U\}.\end{aligned}$$

Set $w=y\wedge z$, $y'=y-w$, $z'=z-w$, $x'=y'-z'$; then the quantity in the outer braces is

$$\begin{aligned}&\sup\{f(y): 0\leqq y\leqq x\}+\sup\{-f(z): 0\leqq z\leqq x\}\\ &= \sup\{f(y)-f(z): 0\leqq y,z\leqq x\}\\ &= \sup\{f(y')-f(z'): 0=y'\wedge z'\leqq y'\vee z'\leqq x\}.\end{aligned}$$

But by (A) and (B) we have

$$\|x'\|=\|y'+z'\|=\|y'\vee z'\|\leqq\|x\|,$$

so the sup above becomes

$$\sup\{f(x'): x'^++x'^-\leqq x\}.$$

Now taking the sup on x in $K\cap U$

$$\begin{aligned}\|f^++f^-\| &= \sup\{f(x'): x'^++x'^-\in K\cap U\}\\ &= \sup\{f(x'): x'\in U\}=\|f\|.\end{aligned}$$

This says that V^* satisfies (A′) which is equivalent to (A) by (9a).

If A is a subset of V^* which has an upper bound b_0, let $B=\{b: A\leqq b\leqq b_0\}$, and if $b\in B$ let $B_b=B\cap(b-K^+)$. Then intervals $a\leqq f\leqq b_0$ are norm-bounded, therefore compact in the w^*-topology, so $B=\bigcap_{a\in A}(a+K^+)\cap(b_0-K^+)$ is w^*-compact. Because V^* is a lattice, the B_b have non-empty finite intersections and are themselves w^*-closed; hence $\bigcap_{b\in B}B_b$ contains some point b_1. Then $A\leqq b_1\leqq B$, but if $b_2\in V^*$ and

$A \leqq b_2$, then $b_0 \wedge b_2 \in B$ so $b_1 \leqq b_2$; that is, b_1 is the least upper bound of A.

Theorem 1. *If V is an* (AB)*-lattice, then V^* and V^{**} are boundedly complete* (AB)*-lattices. The natural mapping Q of V into V^{**} is not only a linear isometry into V^{**} but is also a lattice isomorphism; that is, $Q(x \vee y) = Qx \vee Qy$.*

Proof. Q preserves order, so $Q(x \vee y) \geqq Qx \vee Qy$. To prove equality for all x, y it suffices to prove it for the case where $x \wedge y = 0$.

By (6f) raised one space we have, for each $f \geqq 0$ in V^*,

$$(Qx \vee Qy)(f) = \sup\{g(x) + h(y) \colon g, h \geqq 0 \text{ and } g + h = f\}.$$

But (3) asserts that there exist $g, h \geqq 0$ with $g + h = f$ and $g(x) = f(x)$, $h(y) = f(y)$, so this gives

$$(Qx \vee Qy)(f) \geqq f(x) + f(y) = f(x+y) = f(x \vee y) = [Q(x \vee y)](f).$$

This proves that $Qx \vee Qy = Q(x \vee y)$.

Corollary 1. *Every* (AB)*-lattice has a norm-completion which is also an* (AB)*-lattice.*

Definition 6. (a) A Banach lattice is called an (AM)-*space* (abstract m-space) if the norm and order in the space are related by the conditions (A) and

(M) If $x, y \geqq 0$ then $\|x \vee y\| = \|x\| \vee \|y\|$.

(b) A Banach lattice is an (AL)-*space* (abstract Lebesgue space) if order and norm are related by (A) and

(L) If $x, y \geqq 0$ then $\|x + y\| = \|x\| + \|y\|$.

Clearly either (M) or (L) implies (B), so every (AM)-space, and every (AL)-space, is an (AB)-lattice.

Theorem 2 [Kakutani 2]. *The conjugate space of an* (AM)*-space is an* (AL)*-space. The conjugate space of an* (AL)*-space, and, therefore, the second conjugate space of an* (AM)*-space, is a boundedly complete vector lattice which is an* (AM)*-space whose positive cone has an interior point u such that the unit ball is the order interval $\{x \colon -u \leqq x \leqq u\}$.*

[More can be added to this with the proof of Kakutani's Representation Theorem 2, 2.]

Proof. In the presence of (A) and (B) (which follows from either (M) or (L)), it suffices to check (L) or (M) in V^* for positive elements. Then all that is not immediately verifiable here is the existence of u in V^* if V is an (AL)-space. But the positive cone C in an (AL)-space is a maximal cone in which the norm is additive; hence there is u in V^* such that $C=W(u)$. Since for f in U^π, $\|f^+\| \vee \|f^-\| = \|f\|$, and $\|x\| = u(x) \geqq f(x) \geqq -u(x)$ if $x \in C$, $u \geqq f^+ \geqq f \geqq -f^- \geqq -u$ whenever $\|f\| \leqq 1$. Conversely, $u \geqq f \geqq -u$ implies $u \geqq f^+$ and $u \geqq f^-$, so $\|u\| \geqq \|f^+\| \vee \|f^-\| = \|f\|$.

(11) Suppose that V is a normed vector lattice. (a) If every order interval of V is contained in a ball of V, then $V^* \subseteqq V'$. (b) If f in V' implies that f is bounded on $C \cap U$, the positive part of the unit ball (in particular, if each ball is contained in some order interval) then $V' \subseteqq V^*$. (See Klee [9] for improvements on these and later references.)

(12) Let V be a vector lattice with a cone C which has a core point u; let $\|x\|_u = \inf\{\lambda : \lambda u \geqq x \geqq -\lambda u\}$. Then: (a) This function is a norm in V. (b) This norm in V makes V an (AM)-space if the space is complete. (c) The unit ball is $(u-C) \cap (-u+C)$. Hence (d) a linear functional f on V is continuous if and only if it is bounded on order intervals. (e) If v is another core point of C, then $\| \ \|_u$ and $\| \ \|_v$ are isomorphic, that is, they define the same topology in V.

(13) Properties (A) and (B) are shared by all the common Banach function lattices, $L^p(\mu)$, $p \geqq 1$, $C(S)$, $C_0(S)$. (M) and (L) are very special properties; to show this we give the next section some representation theorems of Kreĭn and Kreĭn [1,2] and Kakutani [2].

The books of Peressini and Schaefer [1] give much more information on ordered linear topological spaces and ordered normed spaces.

§ 2. Linear Sublattices of Continuous Function Spaces

We give in this section Kakutani's characterization [2] of closed sublattices of continuous function spaces; they are isometric to (AM)-spaces. We give first conditions identifying the closed vector sublattices of a given $C(S)$ in terms of the linear relations connecting the values of the functions at different points.

Definition 1. (a) If S is a set, call $S^2 \times R^2$ *the set of (two-point) linear relations over* S and denote it by Λ. Then Λ^+, *the set of non-negative linear relations over* S, is the subset of those (s,s',r,r') in Λ such that $rr' \geqq 0$. (b) If A is a set of real-valued functions on S define $\Lambda(A)$, *the set of linear relations satisfied by* A, to be

$$\{(s,s',r,r') : r x(s) = r' x(s') \text{ for all } x \text{ in } A\},$$

let $\Lambda_0(A) = \{(s,s') \times R^2 : x(s) = 0 = x(s') \text{ for all } x \text{ in } A\}$, and let $\Lambda^+(A) = \Lambda(A) \cap \Lambda^+$. (c) Dually, if Λ' is a subset of Λ and A a subset of real functions on S, then $A(\Lambda')$ is the set of all those functions x in A such that $r' x(s') = r x(s)$ for all (s,s',r,r') in Λ'.

(1) Let A be a subset of $C(S)$, where S is a compact Hausdorff space. Then: (a) If $\Lambda_{ss'}(A)=\{(r,r'):(s,s',r,r')\in\Lambda(A)\}$, then for each (s,s') in S^2 the set $\Lambda_{ss'}(A)$ is a linear subspace of R^2. (b) If $(s,s',r,r')\in\Lambda(A)$, so is (s',s,r',r). (c) If (s,s',r,r') and $(s',s'',t,t')\in\Lambda(A)$, so is $(s,s'',rt,r't')$. (d) If $A'\subseteqq A$, then $\Lambda(A')\supseteqq\Lambda(A)$. (e) Λ (linear hull of A) $=\Lambda(A)$. (f) Λ (closure of A) $=\Lambda(A)$. (g) Λ (smallest linear sublattice of $C(S)$ containing A) $=\Lambda^+(A)\cup\Lambda_0(A)$.

(2) Let $C=C(S)$, S compact Hausdorff; then: (a) If $\Lambda'\subseteqq\Lambda''\subseteqq\Lambda$, then $C(\Lambda')\supseteqq C(\Lambda'')\supseteqq C(\Lambda)=\{0\}$. (b) If $\Lambda'\subseteqq\Lambda$, then $\Lambda(C(\Lambda'))\supseteqq\Lambda'$. (c) If $\Lambda'\subseteqq\Lambda$, then $C(\Lambda')$ is a closed linear subspace of C. If $\Lambda'\subseteqq\Lambda^+$, then $C(\Lambda')$ is a closed linear sublattice of C.

Lemma 1. *If $A\subseteqq C=C(S)$, S compact Hausdorff, then $C(\Lambda^+(A))$ is L_0, the smallest closed linear sublattice of C which contains A.*

Proof. $C(\Lambda^+(A))\supseteqq C(\Lambda(A))\supseteqq A$, and $C(\Lambda^+(A))$ is a closed linear sublattice of C, so $L_0\subseteqq C(\Lambda^+(A))$. To show that each x in $C(\Lambda^+(A))$ can be matched with an element of L_0 we begin with a less ambitious approximation.

(a) Given x in $C(\Lambda^+(A))$, for each s and s' in S there is a y in L_0 such that $y(s)=x(s)$ and $y(s')=x(s')$.

Let F be defined from C into R^2 by $Fz=(z(s),z(s'))$. Since F is linear, (a) is equivalent to the restriction $F(C(\Lambda^+(A)))\subseteqq F(L_0)$. If $F(A)$ contains two linearly independent elements, then $F(L_0)=R^2$. $F(A)$ is contained in a line $\lambda=\{(r,r'):pr-qr'=0\}$ if and only if (s,s',p,q) is in $\Lambda(A)$. If $pq<0$ and z exists in A with $Fz\neq(0,0)$, then $z(s)z(s')<0$ also, so $F(z^+)$ and $F(z^-)$ are linearly independent and $F(L_0)=R^2$. If $pq\geqq 0$, then $(s,s',p,q)\in\Lambda^+(A)$, so Fx is in λ too. If there is a z in A with $Fz\neq(0,0)$, then λ is the line through Fz and the origin, so $Fx\in F(L_0)$. If $F(A)=\{(0,0)\}$, then Fx is on all lines through $(0,0)$, so $Fx=(0,0)$. This proves (a); to continue we prove:

(b) If $x\in C(\Lambda^+(A))$ and $\varepsilon>0$, then there is a z in L_0 such that $\|x-z\|<\varepsilon$.

First choose s' in S; by (a) for each s'' in S there is a y in L_0 such that $y(s')=x(s')$ and $y(s'')=x(s'')$. Hence there is a neighborhood V of s'' in S in which $y(s)>x(s)-\varepsilon$. As s'' varies over S this set of neighborhoods covers S; by compactness there exist $s_1,\dots,s_n$ and corresponding neighborhoods $V_1,\dots,V_n$ and elements $y_1,\dots,y_n$ of L_0 such that $\bigcup_{i\leqq n}V_i=S$; if $w=\bigvee_{i\leqq n}y_i$, then $w(s')=x(s')$ and $w(s)\geqq y_i(s)>x(s)-\varepsilon$ for all s in S. Now to each s' in S there is such a w that is never far below x and agrees with x at s'; hence there is a neighborhood V' in which $w(s)<x(s)+\varepsilon$. The corresponding compactness argument and an inf of w_j gives the desired y in L_0 within ε of x.

(b) proves that L_0 is dense in $C(\Lambda^+(A))$. But L_0 is closed, so the theorem is proved.

Corollary 1. *A set A in $C(S)$, S compact Hausdorff, is a closed linear sublattice of $C(S)$ if and only if $A = C(\Lambda^+(A))$.*

An immediate consequence of this characterization of sublattices of $C(S)$ is a result of Stone [1,2].

Theorem 1 [Stone-Weierstrass]. *Let S be a compact Hausdorff space and let A be a subset of $C(S)$ such that A contains a non-zero constant function and enough other functions to distinguish points of S; then $C(S)$ is the smallest closed vector sublattice of $C(S)$ which contains A.*

Proof. As some non-zero constant function is in A, $(s,s',p,q)\in\Lambda(A)$ if and only if $p=q$. But A also contains a y such that $y(s)\neq y(s')$ so $p=0=q$. Then $\Lambda^+(A)=\Lambda(A)=\{(s,s',0,0)\colon (s,s')\in S^2\}$ and $C(\Lambda^+(A))=C(S)$.

To give a more abstract characterization of the closed linear sublattices of $C(S)$ spaces, observe first that every such space is an (AM)-space. Kakutani [2] proved the converse.

Theorem 2. *A Banach lattice V is isometric and (vector and lattice) isomorphic to a closed linear sublattice of some $C(S)$, S compact Hausdorff, if and only if V is an* (AM)*-space.*

This is related to a theorem of M.G. and S.G. Kreĭn [1 and 2].

Theorem 3. *If V is a vector lattice, then there is a linear and lattice isomorphism T of V onto some $C(S)$, S compact Hausdorff, if and only if there is some core point u of K such that, under the norm $\|x\|_u = \inf\{t\colon -tu\leqq x\leqq tu\}$, V is a complete normed space.*

§ 1, (12c) reduces this and Theorem 1, 2 reduces Theorem 2 to special cases of

Theorem 4. *V is an* (AM)*-space whose cone C has an interior point u such that the unit ball of V is the set $\{x\colon -u\leqq x\leqq u\}$ if and only if there is a compact Hausdorff space S and a linear isometry and lattice homomorphism T of V onto $C(S)$ such that $Tu=e$, the constantly one function on S. S may be taken to be the w^*-topologized set of extreme points of the positive face of the unit sphere of V^*.*

Proof. Each $C(S)$ is an (AM)-space and its unit ball is $\{x\colon -e\leqq x\leqq e\}$; every linear isometry which is also a lattice isomorphism carries these properties with it.

If, on the other hand, V is an (AM)-space, observe first that if an f in V^* is an extreme point of U^π, then either f^+ or f^- is 0, for otherwise, by (L) in V^*, $f=(\|f^+\|/\|f\|)f^+ + (\|f^-\|/\|f\|)(-f^-)$ is in an open segment of U^π. We can restrict attention, therefore to S, the set of extreme points of $K=C^+\cap\{f:\|f\|=1\}=U^\pi\cap\{f:f(u)=1\}$. But "$s$ extreme in K" is equivalent to "s indecomposable in C^+"; by 1, (4), s is a lattice homomorphism of V such that $s(u)=1$. Hence S is a w^*-closed subset of U^π and therefore is compact.

To show that the minimal facets of U^π contain but one point, consider s, s' in S; then $u\in W(s)$ and $u\notin W(-s')$, so no minimal facet contains points of both S and $-S$. If $s\neq s'$, there is y in V such that $s(y)\neq s'(y)$; setting $u=y^+$ or y^-, there is a $z\geqq 0$ with $s(z)\neq s'(z)$; say, for example, that $s(z)>s'(z)\geqq 0$. Then set $w=(z/s(z))\wedge u$ to see that $1=s(w)>s'(w)$, so $w\in W(s)$ while $w\notin W(s')$; that is, s and s' are not in the same minimal facet of U^π.

We now know, by Theorem V, 4, 1 (c) applied to $S\cup -S$, that V is isometric under T, defined by $Tx(s)=s(x)$ for all x in V, to a completely regular subspace of $C(S)$ over the compact space S. But $\vee$ in $C(S)$ is computed pointwise and each s is a lattice homomorphism, so T is a lattice isomorphism and linear isometry of V into $C(S)$; by the Stone-Weierstrass theorem, $T(V)=C(S)$.

Corollary 2. *If the cone C in a Banach lattice V has an interior point u such that the order interval $\{x: -u\leqq x\leqq u\}$ is contained in some ball, then V is isomorphic, as a lattice and a linear normed space, to some $C(S)$, S compact Hausdorff.*

Proof. The hypotheses assert that the norm $\|x\|_u=\inf\{\lambda: -\lambda u\leqq x \leqq\lambda u\}$ is isomorphic to the original norm in V; under the new norm V satisfies the hypotheses of Theorem 4.

(3) Examples of (AM)-spaces not defined initially as spaces of continuous function on compact Hausdorff spaces are: (a) $m(S)$ and $c_0(S)$, S any index set; $m(S)$ is mapped onto the space of continuous functions on the Stone-Čech compactification of S when S is regarded as a discrete space. (b) If all V_s are (AM)-spaces, then $m(S, V_s)$ is also an (AM)-space. (c) If μ is a measure, then $M(\mu)$, the space of bounded, real, μ-measurable functions is an (AM)-space.

(4) It generally happens that the representation of an (AM)-space V through a representation of its second conjugate space is exceedingly redundant; that is, $\Lambda(T(V))$ contains many relations $(s,s',1,1)$. (a) For example, if $V=c_0(S)$, V^{**} is essentially $m(S)$, which is represented on the S$-$Č compactification $\Omega(S)$. But if t and t' are in $\Omega(S)\setminus S$, $(t,t',1,1)$ is in $\Lambda(T(c_0(S)))$; that is, $c_0(S)$ can be represented comfortably on the one-point compactification, $\alpha(S)$. $\alpha(S)$ is homeomorphic to the set of extreme points of $K^+\cap U^\pi$ in $c_0(S)^*$, or, passing to the isometric space $l^1(S)$, $\alpha(S)$ is the set consisting of 0 and the positive basis vectors δ_s under coordinatewise convergence. (b) In general, if the (AM)-space V is represented as a subset $Q(V)$ in V^{**} by continuous functions on

the set S' of positive extreme points of the unit ball in V^{***}, then the mapping Q^* of V^{***} onto V^* carries S' onto a compact space S of elements of $K^+ \cap U^\pi$ in V^*. V is then represented as a closed vector sublattice of $C(S)$, and Kakutani [2] shows that S is the closure of the set of extreme points of the positive face of U^π.

§ 3. Monotone Projections and Extensions

We shall consider in this section a linear space V ordered, as in I, § 6, by means of a wedge W. The problem discussed is a generalization of the monotone and the dominated (or Hahn-Banach) extension theorem of I, § 6: What are the conditions on V under which these extension properties hold not for real-valued functions but for functions with values in V? It turns out that this depends heavily on bounded completeness of the lattice of V, but does not depend at all on uniqueness of "least" upper bounds.

Theorem 1. *Let W be a wedge in a linear space V; then the first five conditions below are equivalent to each other; any one of them implies the last and implies that the wedge W is lineally closed (that is, that the intersection of W with every line in V is a closed set in the line). If W is lineally closed and* (HB) *holds, then the other conditions also hold.*

(INT) (Interpolation property). For each pair of non-empty sets $A \leqq B$ (that is, such that $a \leqq b$ for each a in A and b in B) in V, there is an element v between A and B, $A \leqq v \leqq B$.

(LUB) W is a fully minihedral wedge, that is, every set in V which has an upper bound has a least upper bound.

(IME) ("Into" monotone extension property). For each OLS Y (Def. I, 6, 2) with wedge W', and each linear subspace X of Y such that for each y in Y, $y+X$ meets W' if and only if $y+X$ meets $-W'$, every linear monotone f from X into V has a linear monotone extension F from Y into V.

(MP) (Monotone projection property). For each OLS Y containing V with wedge W' in Y such that $W' \cap V = W$ and such that $y+V$ meets W' if and only if $y+V$ meets $-W'$, there is a monotone projection P (that is, a monotone idempotent linear operator) carrying Y onto V.

(FME) ("From" monotone extension property). If Y and V satisfy the hypotheses of (MP), then every monotone linear f from V into an OLS X has a monotone linear extension F from Y into X.

(HB) (Hahn-Banach extension property). For each linear space Y and sublinear function p from Y into V, each linear f defined on a linear subspace X of Y and dominated there by p has a linear extension F defined from Y into V and dominated everywhere by p.

Proof. To see that (INT) implies (LUB) let B be the set of all upper bounds of A; then the v from (INT) is a least upper bound for A. The proof of Theorem I, 6, 1 is precisely the proof that (LUB) implies (IME). (IME) specializes to (MP) when $X=V$ and f is the identity in V. (FME) follows from (MP) by setting $F=f\circ P$, and it implies (MP) by specialization. Using $V\times Y$ as $R\times Y$ was used in I, 6, (8), (MP) implies (HB).

It remains to be shown that either (MP), or (HB) combined with lineal closure of W, suffices to give (INT). (This final implication is due to Silverman and Yen). Take non-empty sets $A\leqq B$ in V and subtract some element a' of A to get, with no loss of generality, $0\in A\leqq B\subseteqq W$. Unless $W=V$, in which case (INT) holds trivially, there is an element w in W such that $-w\notin W$; add this w to any b of B to get a point b_0 of $B+W$ such that $-b_0\notin W$ and $b_0\geq A$.

Step 1. Define $A_0=\{a:0\leq a\leq B\}$, $B''=\{b':b'\geq A_0\}$, and $B_0=B''\cap(b_0-W)$. Let V' be a larger linear space containing V and one y_0 not in V, and let W' be the smallest cone in V' containing W, y_0-A_0, and B_0-y_0. Then $W'\cap V=W$.

If it is assumed that (MP) holds, there is a monotone projection P of Y' (ordered by W') onto V; let $v=Py_0$; then $A\leq v\leq B$.

If (HB) and lineal closure hold for W, then define $p(v)=[\inf\{t:tb_0-v\in W'\}]b_0$; then $p(y)<[+\infty]b_0$ on a wedge U in Y and p is subadditive and positive-homogeneous in U. Let $Y=U+(-U)$ and $X=Y\cap V$; then Y is linear and $b_0\in U$ with $p(b_0)=b_0$. $A_0\cup B_0\subseteq X$ because $0\leq p(a)\leq p(b)\leq b_0$. $y_0\in U$ because $b_0-y_0\in W'$ so $p(y_0)\leq b_0$. $y_0\in W'$ so $b_0+y_0\in W'$ or $p(-y_0)\leq b_0$ also. Lineal closure of W implies that of $W\cap X$, so $p(x)\geq x$ for all x in X. Therefore there is a linear F from Y into V such that $-p(-y)\leq F(y)\leq p(y)$ for all y in Y and $F(x)=x$ for all x in X.

Set $v_0=F(y_0)$. Then if $b\in B_0$, also $-y_0+b\in W'$, $y_0-b\in -W'$, and $p(y_0-b)\leq 0$, so $F(y_0-b)\leq 0$; that is, $v_0\leq B_0$. Also, if $a\in A$, then $y_0-a\in W'$, so $a-y_0\in -W'$ and $0\geq p(a-y_0)\geq F(a-y_0)=a-v_0$, so $A_0\leq v_0$. From this we see that $A_0\leq v_0\leq B_0$ and also $v_0\in B_0$.

Step 2. Consider b_0 and any b in B'' and take $b'=b+b_0$. Then $-b'\notin W$ and $B'=B''\cap(b'-W)\supseteq B_0\neq\emptyset$. Step 1 shows that there is v' such that $A_0\leq v'\leq B'$, so $v'\in B''$ and $v'\leq b$ and $v'\leq b_0$; hence $v'\in B_0$, so $v_0\leq v'\leq b$. Hence $0\leq v_0\leq B$, so $v_0\in A_0\cap B''$. [Each element u of this set satisfies $u\leq v_0\leq u$.]

Step 3. If $a\in A$, let $d=b_0-a$; then $d\in W$ and $b_0=a+d\in W$. Define $D=B+d$, $d_0=b_0+d$, $C=\{c:0\leq c\leq D\}$, and $D''=\{d:C_0\leq d\}$. By Step 2 there is w in $C\cap D''$; let $u=w-d$.

Then $A_0+d\subseteq C\leq w$ so $A_0\leq u$, or $u\in B''$ and $0\leq u$.

Also $w<D''\supseteq D=B+d$, so $0\leq u\leq B$ or $u\in A_0$. That is, $u\in A_0\cap B''$ as does v_0, so $u\leq v_0\leq u$. But also $u\geq a$, so $v_0\geq u\geq a$. This holds for all a in A so $A\leq v_0\leq B$. (In fact, $A_0\cap B''$ consists entirely of lower bounds of B which are upper bounds of A.)

(1) Some examples of spaces with these properties are: (a) $m(S)$ or $l^p(S)$, $p>0$, S any index set, with the cone W of nowhere-negative functions. (b) $L^p(\mu)$, μ any measure, $p>0$, with the wedge W of almost everywhere non-negative functions. (c) $M(\mu)$, μ finite on the whole space. (This is a sublattice of $L^1(\mu)$.)

(2) (a) If S is a compact Hausdorff space, then $C(S)$ has (LUB); that is, is a boundedly complete vector lattice, if and only if S is extremally disconnected; that is, the closure of every open subset of S is open. [See Goodner.] [If A is in $C(S)$ with upper bound b, for each s let $f(s)=\sup\{x(s): x \text{ in } A\}$. Then f is lower semicontinuous, so for each real t the set $G_t=\{s; f(s)>t\}$ is open; let H_t be the closure of G_t. Define $g(s)=\sup\{t: s\in H_t\}$. Then $g\geq f$ and $\{s: g(s)\geq u\}=H_u$, which is both open and closed since S is extremally disconnected. Hence $g\in C(S)$ and is the supremum of A.] (b) If Y is a normed linear space, if X is a linear subspace of Y and if S is an extremally disconnected, compact Hausdorff space, let $p(y)=\|y\|e$, where e is the constantly-one function on S. If f is a continuous linear function from X into $C(S)$, then $\|f\|p(x)\geqq f(x)$ for all x in X and $\|f\|p$ is sublinear. By (a) and Theorem 1, $C(S)$ has (HB), so f has an extension F defined from Y into $C(S)$ with $F(y)\leqq\|f\|p(y)$ for all y. Hence $\|F\|=\|f\|$ and, by V, 5, (9c), $C(S)$ has property $\mathfrak{P}_1$; this completes the proof of Theorem V, 4, 3, (b) implies (a).

(3) V, §2 discussed another kind of generalization of the Hahn-Banach theorem in which the values of the functions are real but invariance under a family of linear operators is required. For a generalization in both directions simultaneously, see R. J. Silverman [1, 2].

§ 4. Special Properties of (AL)-Spaces

It is already known that the conjugate of an (AB)-lattice is a boundedly complete (AB)-lattice; an (AL)-space, even if not a conjugate space, has this property.

Theorem 1. *In an* (AL)-*space* V, *every set bounded above has a least upper bound.*

Proof. If $A_0\leqq b$, we may suppose, by subtracting an a_0 of A_0, that $0\in A_0\leqq b$. Let A_1 be the set of suprema of finite subsets of A_0 and let

A be the set of non-negative elements of A_1. Then A is a directed system bounded above by b, so $k = \limsup_{a \in A} \|a\| < \infty$. Let (a_n) be a non-decreasing sequence of elements from A chosen so that $\lim_{n \in \omega} \|a_n\| = k$, and let $x_1 = a_1$, $x_{n+1} = a_{n+1} - a_n$ for $n > 1$. Then $x_i > 0$ so

$$\sum_{i \leqq n} \|x_i\| = \left\| \sum_{i \leqq n} x_i \right\| = \|a_n\| \leqq k;$$

hence $\sum_i x_i$ is an absolutely convergent series in V which must have a sum $b_1 = \lim_{n \in \omega} a_n$. If $a \in A$, then

$$\begin{aligned} k = \|b_1\| &\leqq \|b_1\| + \|(b_1 \vee a) - b_1\| = \|b_1 \vee a\| \\ &= \left\| \left(\lim_{n \in \omega} a_n \right) \vee a \right\| = \left\| \lim_{n \in \omega} (a_n \vee a) \right\| \leqq k. \end{aligned}$$

Hence $\|(b_1 \vee a) - b_1\| = 0$ so $b_1 \vee a = b_1$ if $a \in A$. But each element of A_1 is below an element of A, so b_1 is the least upper bound for A_1, and hence for A_0.

This property was used by Kakutani [1] to find a Boolean algebra of elements of an (AL)-space V as a step in representing V as a space of functions summable with respect to some measure. Call an element v in V a *Freudenthal unit* (or F-unit) for V if $v \wedge x = 0$ implies $x = 0$.

Theorem 2. *For each* (AL)*-space V there is a set S, a Boolean σ-algebra $\mathscr{S}$ of subsets of S, and a measure μ on $\mathscr{S}$ such that V is isometric and isomorphic (both linearly and latticially) to $L^1(\mu)$. In case V has a Freudenthal unit, S may be taken compact Hausdorff, $\mu(S)$ finite (even one), and $\mathscr{S}$ chosen so each of its elements differs by a set of μ-measure zero from an open-and-closed subset of S. If V has no Freundenthal unit, S may be taken as a union (necessarily uncountable) of such compact Hausdorff spaces S_t, $\mathscr{S}$ the family of countable unions of elements of the corresponding $\mathscr{S}_t$, with $\mu\left(\bigcup_{i \in \omega} E_{t_i}\right) = \sum_{i \in \omega} \mu_{t_i}(E_{t_i})$.*

The proof is to be found in Kakutani [1]. In the case where an F-unit v exists, for each set A of non-negative elements of V, Theorem 1 asserts that there is a largest $a' \leqq v$ such that $a' \wedge a = 0$ for all a in A. The family of all such a' is a Boolean σ-algebra, even a complete Boolean algebra, with a measure $\mu'(a') = \|a'\|$. This can be represented as in Stone [1]. In the general case, the family T is a maximal family of positive elements t of V for which $t \wedge t' = 0$ whenever $t \neq t'$. This splits V into an $l^1(T)$-sum of (AL)-spaces $V_t \subseteqq V$, where $x \in V_t$ if and only if $x^+ \wedge t'$ and $x^- \wedge t'$ are 0 for each $t' \neq t$.

(1) Some results of Maharam imply that each (AL)-space V with an F-unit can be represented as an $l^1(\omega)$ sum of a sequence of spaces $L^1(\mu_n)$, where μ_0 is an atomic measure on all subsets of some finite or

countable set S_0, and where each μ_n, $n>0$, is a scalar multiple of product measure on a compact discontinuum $S_n = 2^{E_n}$, where if $n>m$, the cardinal number of $E_n >$ that of $E_m \geqq \aleph_0$. See Day [6] where this is applied to show that *every* (AL)-*space is isomorphic to a rotund space* (VII, § 2).

Consequences of Theorem 2 and the well-known properties of $L^1(\mu)$ spaces are other theorems of Kakutani [1].

Theorem 3. *Every interval* $\{x: a \leqq x \leqq b\}$ *in an* (AL)-*space is w-compact.*

Theorem 4. *Every* (AL)-*space is w-ω-complete.*

Kakutani shows in the proof of Theorem 2 that every separable (AL)-space is isometric and isomorphic with a Banach sublattice of $L^1 = L^1(\mu)$, μ Lebesgue measure on $[0,1]$. Then Theorem 4 follows from the corresponding theorem proved for L^1 by Steinhaus, and Theorem 3 from the corresponding theorem proved for L^1 by Lebesgue. (See Banach, p. 136.)

Theorem 5 (Dunford-Pettis). *If* T_1 *and* T_2 *are wcc linear operators from an* (AL)- [*or* (AM)-] *space* V *into itself, then* $T_1 \circ T_2$ *is norm cc.*

This was proved by Dunford and Pettis for $L^1(\mu)$, μ Lebesgue measure. Generalizations and later references can be found in Bartle, Dunford and Schwartz and in Grothendieck [5]. Once it is known for every (AL)-space, Theorems III, 3, 2 and VI, 1, 2 show that it is true in every (AM)-space.

The proof in $L^1(\mu)$ depends on the result of Dunford-Pettis that *a wcc linear operator from* $L^1(\mu)$ *into itself carries each weakly convergent sequence into a norm-convergent sequence.* This fact, in turn, is largely dependent on the Orlicz-Pettis theorem IV, 1, 1.

A proof of the following generalization, due to Kakutani [2], of the Riesz representation theorem for linear functionals can be found in Halmos [1], Chapter 10.

Theorem 6. *Let* S *be a compact Hausdorff space and let* $\mathscr{S}$ *be the family of all Baire sets in* S*; that is,* $\mathscr{S}$ *is the Boolean σ-algebra of subsets of* S *generated by the compact* G_δ *subsets of* S*. Then each* f *in* $C(S)^*$ *determines a unique countably-additive real function* μ *defined on* $\mathscr{S}$ *such that* $f(x) = \int_S x\, d\mu$ *and* $\|f\| =$ *total variation of* μ.

The construction is essentially that described by Riesz and Nagy, a variant of the Daniell integral.

(2) There have been many generalizations of Riesz's original representation theorem for linear functionals on $C[0,1]$. Gel'fand and Dunford-Pettis gave many representations for linear operators between

familiar function spaces. More recently, Grothendieck [5] and Bartle, Dunford, and Schwartz (see these papers for further references) have discussed *wcc* linear operators from $C(S)$ into a Banach space B. Basically, the B-D-S result is that such a T is representable as $Tx = \int_S x\,d\varphi$, where φ is a vector measure; that is, a function defined on a suitable σ-algebra $\mathscr{S}$ of subsets of S into B such that for each f in B^*, $f\circ\varphi$ is an ordinary real-valued, countably additive set-function on $\mathscr{S}$. B-D-S also show that φ has this property if and only if the image of the unit ball U^π of B^* under the mapping $\Phi(f) = f\circ\varphi$ is a relatively w-compact subset of $ca(\mathscr{S})$, the Banach space of all countably additive, real-valued set functions ψ on $\mathscr{S}$ with $\|\psi\| =$ total variation of ψ.

(3) (a) A corollary of Theorem 6 and of the Lebesgue dominated convergence theorem is: A *sequence* $(x_n, n\in\omega)$ is w-convergent to x in $C(S)$, where S is a compact Hausdorff space, if and only if $\{\|x_n\| : n\in\omega\}$ is bounded and $\lim_{n\in\omega} x_n(s) = x(s)$ for every s in S. [See Banach, p. 224, Th. 8 for metric S.] (b) Šmul'yan [9] gives an examples of a *net* in $C(S)$ which is bounded and pointwise convergent to zero but is not weakly convergent.

4 (a) Grothendieck [5] discovered another unusual property of $m(S)^*$; (M_ω) *Every* w^**-convergent sequence* $(\Xi_i, i\in\omega)$ *is* w*-convergent.* See III, 4, (8). (It suffices to assume that $w^*\text{-}\lim_i \Xi_i = 0$ and then to prove that $\{\Xi_i\}$ is w-ω-relatively compact. Shift attention to the matching measures μ_i on the Stone-Čech compactification $\Omega(S)$ and let $\mu = \sum_{i\in\omega} |\mu_i|/2^i$. Then each μ_i is absolutely continuous with respect to μ, so we can shift attention again to matching f_i in $L_1(\mu)$. If $\{\Xi_i\}$ is not w-ω-compact in $m(S)^*$, then $\{f_i\}$ is not w-ω-compact in $L_1(\mu)$; use the criterion of III, 2, (9a), to see that there exist $\varepsilon > 0$, measurable sets E_j, and indices i_j such that $\int f_{i_j}\,d\mu \geq 8\varepsilon$ for all j in ω. Subsequencing on j, either $\int_{E_j} f_{i_j}^+\,d\mu \geq 4\varepsilon$ or $\int_{E_j} f_{i_j}^-\,d\mu \geq 4\varepsilon$; the first case is typical. Then there exist open-and closed sets $U_j \supseteq E_j$ such that $|\mu_{i_j}|(U_j \setminus E_j) < 2\varepsilon$ so $\mu_{ij}(U_j) \geq 2\varepsilon$. But $\lim_j \mu_i(U_j) = 0$ since $\chi_{U_j} \in C(\Omega(S))$; by subsequencing again and setting $V_1 = U_1$, $V_2 = U_{j_2} \setminus U_1$, etc., we get $\mu_{i_j}(V_j) \geq \varepsilon$ and V_j disjoint open-and-closed subsets of $\Omega(S)$. Then the map T of $m(\omega)$ into $C(\Omega(S))$ for which $Tx = x_n$ in V_n, $Tx = 0$ outside the closure of $\bigcup_{n\in\omega} V_n$, and Tx is defined by continuity elsewhere in $\Omega(S)$, can be used in Cor. II, 2, 1 to say that $\varepsilon \leq \mu_{i_n}(V_n) \leq \sum_{j\in\omega} |\mu_{i_n}(V_j)| \to 0$, a contradiction which tells us that $\{\Xi_i\}$ (and even $\{\Xi_i^+\}$) is a w-ω-relatively compact

set. But each w-convergent subsequence of (Ξ_i) is also w^*-convergent to 0, so (Ξ_i) is w-convergent to 0.)

(b) A generalization and consequence of (a) is also due to Grothendieck [5]. *If S is compact Hausdorff and extremally disconnected, then every w^*-convergent sequence in $C(S)^*$ is w-convergent to the same limit.* (By Theorem V, 4, 3, when $C(S)$ is represented as a subspace of $m(U^\pi)$ there is a projection P of norm one of $m(U^\pi)$ on $C(S)$. P^* carries w^*convergent sequences in $C(S)^*$ to w^*-convergent sequences in $m(U^\pi)^*$.)
(c) The same proof shows that *if B is a Banach space in the family $\mathfrak{P}$* (Def. V, 4, 1), *then w^*- and w-sequential convergence are equivalent in B^*.*
(d) Schaefer [2], among others, has polished Grothendieck's proof to carry through for a $C(S)$ with S quasi-Stonean, that is, the closure of every open F_σ-subset of S is open. This property of S is equivalent to: Every countable bounded set in $C(S)$ has a least upper bound.

Chapter VII. Metric Geometry in Normed Spaces

§ 1. Isometry and the Linear Structure

Banach's book, p. 160, gives a theorem of Mazur and Ulam that *an isometry of one normed space onto another which carries* 0 *to* 0 *is linear.* This is true only for real-linear spaces, and is proved by characterizing the midpoint of a segment in a normed space in terms of the distance function. Using the same proof a slightly stronger result can be attained.

Theorem 1. *In a locally convex linear topological space over the real field the uniformity and the zero point determine the linear structure.*

Most of the proof depends on

Lemma 1. *Let L be an* LCS *and let $\{p_s: s\in S\}$ be a family of continuous pre-norms large enough to separate points of L; for example, this might be the family of all continuous pre-norms in L. For each s let d_s be defined by $d_s(x,y)=p_s(x-y)$. Then for each x_1, x_2 in L the midpoint $x_0=(x_1+x_2)/2$ of the segment from x_1 to x_2 can be found in terms of the set $\{d_s: s\in S\}$ of pre-metrics.*

Proof. To save subscripts, temporarily let d be any one of the d_s; then define $E_1=\{x: x\in L$ and $d(x_1,x)=d(x,x_2)=d(x_1,x_2)/2\}$. If E_n is defined, let $D(E_n)$ be the diameter of E_n measured in terms of the pre-metric d; then set $E_{n+1}=E_n\cap\{x: d(x,y)\leqq D(E_n)/2$ for every y in $E_n\}$. It follows easily that if E_{n+1} is not empty, then $D(E_{n+1})\leqq D(E_n)/2$.

Next it is necessary to show that $x_0\in\bigcap_n E_n$; to this end define T by the formula $Tx=x_1+x_2-x$ for each x in L. Then T is an isometry whose only fixed point is x_0, and $x_0\in E_1$. To prove by induction on n that $T(E_n)\subseteq E_n$ for all n, take x in E_1; then $d(Tx,x_i)=d(x_1+x_2-x,x_i)$ $=d(x_{3-i},x)$, so $Tx\in E_1$. Next suppose y and Ty in E_n and x in E_{n+1}, then

$$d(Tx,y)=d(x_1+x_2-x,y)=d(x_1+x_2-y,x)=d(x,Ty)\leqq D(E_n)/2,$$

so $Tx \in E_{n+1}$, and induction proves $T(E_n) \subseteqq E_n$ for all n. Then for x in E_n

$$d(x, x_0) = d(Tx, x_0) = \frac{d(x, Tx)}{2} \leqq \frac{D(E_n)}{2},$$

so $x_0 \in E_{n+1}$ if $x_0 \in E_n$. This proves $x_0 \in \bigcap_n E_n = E$.

Now consider again the set of all $\{d_s : s \in S\}$. By the preceding argument each d_s determines a set E^s in which x_0 lies; also the d_s-diameter, $D_s(E^s)$, is zero; that is, if $x \in \bigcap_{s \in S} E^s$, then $d_s(x, x_0) = 0$ for all s in S. But it was assumed to begin with that the family of pre-metrics was large enough to separate points; hence x_0 is the only element of $\bigcap_{s \in S} E^s$.

The theorem will be proved if we add to the lemma the following special case of the theorem.

Corollary 1. *Call a one-to-one function T between two locally convex spaces a unimorphy if it carries a separating family $\{d_s : s \in S\}$ of pre-metrics on L to another such family $\{d'_s : s \in S\}$ on L' by the rule $d'_s(Tx, Ty) = d_s(x, y)$. Then each unimorphy T is affine.*

Proof. Given x_1 and x_2 in L, T carries the set E^s (of the lemma) to the corresponding set E'^s in L' constructed in a similar manner from Tx_1 and Tx_2. Hence $(x_1 + x_2)/2$, the only point in all the E^s, is carried by T to $(Tx_1 + Tx_2)/2$, the only point in all of the E'^s; this asserts that T preserves midpoints. As in the proof of Lemma I, 4, 1 it can now be proved that $T(r x_1 + (1-r) x_2) = r\,Tx_1 + (1-r)\,Tx_2$ if r is dyadic rational. The families $\{d_s : s \in S\}$ and $\{d'_s : s \in S\}$ determine locally convex topologies in L and L', respectively, in which T is continous; hence, the above relation holds for all real r, and T is affine.

This completes the proof of the corollary. If also $T0 = 0$, then $T(rx) = T(rx + 0) = T(rx + (1-r)0) = r\,Tx$; hence T is linear.

This proofs applies, of course, to the normed case, where one norm is a separating family of continuous pre-norms.

Charzynski has shown that an isometry of a *finite-dimensional* linear metric space which leaves 0 invariant must be linear. His proof consists of constructing from the given metric a pre-norm which is not identically zero and which is invariant under every isometry which has 0 as a fixed point. The proof then uses the Mazur-Ulam result and induction on the dimension of the space.

Mankiewicz gives a generalization of the Mazur-Ulam theorem: *Let V be an open connected subset of a normed space N and let T be an isometry of V onto an open subset of normed space N'; then T can be extended (uniquely) to an affine isometry of N onto N'.* [The proof begins by proving T affine in some ball about each point of V; connectedness makes consistent extension possible. A simple example in a two-dimensional l^1 space shows that connectedness of V is needed.]

§ 2. Rotundity and Smoothness

If a convex set K has interior points, then it may happen that no open segment in K contains a boundary point of K. Such a set has been called "strictly convex" in many papers; here it shall be called rotund. To simplify the discussion it shall be assumed that the convex body is also bounded and symmetric about 0; then it may be taken to be the unit ball of a normed linear space.

In this section N will be a normed space, B its completion, U the unit ball, Σ the surface of $U = \{x: \|x\| = 1\}$, U^π the unit ball of $B^* = N^*$, and Σ' the surface of U^π. Much of what follows can be found in Šmul'yan [3, 6, 7], Klee [6], or Day [1 to 5].

Definition 1. N, or U, is *rotund* (R) if every open segment in U is disjoint from Σ. N, or U, is *smooth* (S) if at each point of Σ there is only one supporting hyperplane of U.

(1) There are a number of other properties of N easily shown to be equivalent to (R): (R_1) Every point of Σ is an extreme point of U. (R_2) Every point of Σ is an exposed point of U. (R_3) Every hyperplane of support of U touches Σ in at most one point (Ruston [3]). (R_4) Supporting hyperplanes to U at distinct points of Σ are distinct. (R_5) If W is a wedge in N in which norm is additive, then W is a halfray $\{tx: t \geq 0\}$. (Kreĭn calls such a U *strictly normalized;* he uses the property in work on the moment problem; see Ahiezer and Kreĭn.) (R_6) In every two-dimensional subspace L of N the unit ball $L \cap U$ is rotund.

(2) Properties equivalent to smoothness are:

(S_1) In every two-dimensional linear subspace L of N the unit ball $L \cap U$ is smooth. [The proof that (S_1) implies (S) requires the Hahn-Banach theorem.]

(G) The norm functional in N has a Gateaux differential at each point of Σ; that is,

$$\text{(i)} \qquad G(x,h) = \lim_{t \to 0} \left(\frac{\|x + th\| - \|x\|}{t} \right)$$

exists for each x in Σ and h in N. (Because $\|\ \|$ is a convex function, this implies that $G(x,h)$ is a linear function of h.)

(3) (a) If N^* is (R) [or (S)], then N is (S) [or (R)]. (b) Hence, if B is reflexive, B is (R) [or (S)] if and only if B^* is (S) [or (R)]. (c) If $l^1(\omega)$ is renormed as in 4, (2), it will be rotund, but $l^1(\omega)^*$ is like $m(\omega)$ which, by 4, (1), has no isomorphic smooth norm. Hence the duality of (b)

requires reflexivity. (d) In the absence of reflexivity there is a substitute duality for going upstream: U^π *is* (R) [or (S)] *if and only if every two-dimensional factor space of* B *is* (S) [or (R)]. [This depends on the isometries described in Cor. II, 1, 1 and, of course, the two-dimensional case of (b). The uniformization of this fact is the crux of the argument of Day [4] sketched in (10a).] (e) Hence, by (a), for B to be (R) or (S) it *suffices* that every two-dimensional factor space of B have the same property. The example of (c) shows that rotundity of B need not imply rotundity of every factor space. Troyanski [3] has given a corresponding example for smoothness.

Next we consider some functions, *moduli of rotundity*, which measure the minimum depth of a segment in U subject to various constraints. To say that one of these moduli for a space N is positive has a strong effect on the geometrical and other properties of N and U.

Definition 2. In the formulas below $\Delta(x,y)=1-\|(x+y)/2\|$ is the depth of the midpoint of the segment $[x,y]$ below Σ and the infimum is calculated subject to the general constraint that x and y are both in U.

$$\delta(\varepsilon)=\inf\{\Delta(x,y)\colon \|x-y\|\geq\varepsilon\}.$$

If β is in Σ', then

$$\delta(\varepsilon,\beta)=\inf\{\Delta(x,y)\colon |\beta(x-y)|\geq\varepsilon\}.$$

If x is in Σ, then

$$\delta(x,\varepsilon)=\inf\{\Delta(x,y)\colon \|x-y\|\geq\varepsilon\}.$$

If x is in Σ and β in Σ', then

$$\delta(x,\varepsilon,\beta)=\inf\{\Delta(x,y)\colon |\beta(x-y)|\geq\varepsilon\}.$$

If z is in Σ then

$$\delta(\varepsilon,\to z)=\inf\{\Delta(x,x+\lambda z)\colon \|\lambda z\|\geq\varepsilon\}.$$

Say that N, or U, is

(UR) uniformly rotund if $\delta(\varepsilon)>0$ when $0<\varepsilon\leq 2$;

($\mathrm{UR}^{A'}$) when A' is a subset of Σ' and $\delta(\varepsilon,\beta)>0$ for each β in A' and each ε with $0<\varepsilon\leq 2$; in particular, we have

($\mathrm{UR}^{\Sigma'}$) = (wUR), *weakly uniformly rotund* [Šmul'yan 6];

(UR_A) when A is a subset of Σ and $\delta(\varepsilon,\to z)>0$ for each z in A and each ε, $0<\varepsilon\leq 2$; in particular,

(UR_Σ) is *directionally uniformly rotund* (the *uniformly convex in every direction* of Garkavi and Day, James, and Swaminathan);

(LUR) *locally uniformly rotund* if $\delta(x,\varepsilon)>0$ for each x in Σ and ε with $0<\varepsilon\leq 2$;

(wLUR) *weakly locally uniformly rotund* or $(\mathrm{LUR}^{\Sigma'})$ if $\delta(x,\varepsilon,\beta)>0$ for each x in Σ, β in Σ, and ε with $0<\varepsilon<2$;

(NQ) *inquadrate* if there exists ε with $0<\varepsilon<2$ such that $\delta(\varepsilon)>0$; this is the *uniformly non-square* of James [6] which is discussed in § 4. (See Defn. 4, 3.)

(4) Let A' be a subset of Σ' which is total over N; then (a) (UR)⇒(NQ). (b) (UR)⇒(LUR)⇒(R). (c) (UR)⇒$(\mathrm{UR}^{\Sigma'})$⇒$(\mathrm{UR}^{A'})$⇒(UR_{Σ})⇒(R).

(5) Some reformulations of (UR) are: (UR_{ω}) If $(x_n, n\in\omega)$ and $(y_n, n\in\omega)$ are sequences in U such that $\lim_{n\in\omega}\|x_n+y_n\|=2$, then $\lim_{n\in\omega}\|x_n-y_n\|=0$. (UR_2) (Kračkovskiĭ and Vinogradov). For $0<\varepsilon<2$ there exists $\delta'(\varepsilon)>0$ such that if H_x and H_y are hyperplanes of support to U at the points x and y of Σ, and if $\|x-y\|\ge\varepsilon$, then $\|H_x\cap H_y\|\ge 1+\delta'(\varepsilon)$. [This can also be formulated in terms of sequences. One can show also that $\delta(\varepsilon)\ge\delta'(\varepsilon/2)$ and $\delta'(\varepsilon)\ge\delta(\varepsilon/2)$.] $(U\nu)$ (Ruston [3]) for $0<\varepsilon\le 2$ there exists $\delta''(\varepsilon)>0$ such that β in Σ and diam $\nu(\beta,\delta)\ge\varepsilon$ imply $\delta\ge\delta''(\varepsilon)$. (See (ν) below.)

There are other rotundity conditions which are not easily formulated in terms of moduli: N, or U, is

(kR) *k-rotund*, where $1<k\in\omega$, if each sequence $(x_n, n\in\omega)$ such that
$$\lim_{n_1,\ldots,n_k\to\infty}\Big\|\sum_{i\le k} x_{n_i}\Big\| = k \text{ is a Cauchy sequence.}$$

(K) If K is a convex set in N, then diam $(K\cap tU)$, the diameter of $K\cap tU$, tends to 0 as t decreases towards the distance from 0 to K.

(K_{ω}) If K is a convex set in N and (x_n) is a sequence of elements of K such that $\|x_n\|$ tends to the distance from 0 to K, then (x_n) is a Cauchy sequence.

(ν) If $\beta\in\Sigma'$ and $\nu(\beta,\delta)=\{x\colon x\in U \text{ and } \beta(x)\ge 1-\delta\}$, then diam $\nu(\beta,\delta)$ tends to 0 as δ decreases to 0. [Thin nibbles are small nibbles.] [This condition can also be reformulated in terms of sequences.]

$(\mathrm{L}\nu)$ Same as (ν) for those β which attain their supremum on U.

(H) If (x_n) is a sequence which converges weakly to x and if $\|x_n\|$ tends to $\|x\|$, then $\|x_n-x\|$ tends to 0; that is, for sequences in Σ weak and norm convergence agree.

(ρ) the completion of N is reflexive.

(6) (a) (K)⇔(K_{ω})⇔(ν)⇒(completion of N is (J) of III, § 2)⇒(ρ). (b) (UR)⇒(2R)⇒ ··· (kR)⇒$((k+1)\mathrm{R})$⇒(K)⇔(ν)⇒$(\mathrm{L}\nu)$⇒(H). (c) (LUR) ⇒$(\mathrm{L}\nu)$⇒(R_3).

(7) Let $B=P_2\, l^{q_i}$, where $1<q_i<\infty$ for each i in ω. [See II, 2, (11) and also (11) below.] (a) Day [2] shows that B is (UR) if there exist numbers $1<m\le q_i\le M<\infty$ for all i in ω, and Day [1] shows that otherwise B is not isomorphic to a (UR) space. (b) Fan and Glicksberg show that B is (kR) for all $k>1$ and all allowed choices of q_i. (c) Lovaglia

shows that B is (LUR). (d) Day, James, and Swaminathan show that $P_2 l^{q_i}$ is (UR_Σ); (11) even asserts that it is $(\mathrm{UR}^{\Sigma'})$. This shows that even the combination of conditions (LUR), (kR), and $(\mathrm{UR}^{\Sigma'})$ is not sufficient to imply that B is (UR) or even isomorphic to a (UR) space. See Th. 4, 4.

(8) (a) (H) is a property of all finite-dimensional spaces so does not imply (R). (b) The condition (H) is used by Kadec [4] as a half-way stage in his proof that every separable N is renormable to be (LUR). The proof of Troyanski [2] that every (WCG) space is ⟨LUR⟩ also seems to work past this property. V.D. Mil'man [11, Chap. 1, § 2], gives Kadec's proof of ⟨LUR⟩ and also generalizes H to a property $(\mathrm{H}^{A'})$ by requiring convergence for all β in A' rather than requiring full weak convergence. (c) If N is (K), it is possible to use Cor. II, 4, 2 to prove directly, instead of through (J), that N is (ν). (d) Lovaglia showed that $l^1(\omega)$ is isomorphic to an (LUR) space, so (LUR) does not imply reflexivity. (e) (NQ) implies reflexivity; see James [6] or Theorem 4, 4. (NQ) does not imply (R); let N be the two-dimensional space with hexagonal unit ball.

Definition 3. In the limit (i) defining Gateaux differentiability of the norm it is possible to add uniformity conditions in several ways: Say that N, or U, is

(UG) *uniformly Gateaux differentiable* if for each h in Σ the limit in (i) is attained uniformly for x in Σ;

(F) *Fréchet differentiable* if for each x in Σ the limit in (i) is attained uniformly for h in Σ;

(UF) *uniformly Fréchet differentiable* if the limit in (i) is attained uniformly for x in Σ and h in Σ;

(US) uniformly smooth if for each $\eta > 0$ there exists $\varepsilon(\eta) > 0$ such that if $\|x-y\| \leq \varepsilon$, then $\|x+y\| \geq \|x\| + \|y\| - \eta \|x-y\|$.

(9) (a) These differentiation conditions can also be defined in terms of moduli of smoothness related to them as (US) is related to (UF); see V. D. Mil'man [11]. (b) (US)$\Leftrightarrow$(UF)$\Rightarrow$(F)$\Rightarrow$(G)$\Leftrightarrow$(S). (c) (UF) $\Rightarrow$(UG)$\Rightarrow$(G).

There is a good deal known about duality of these rotundity and smoothness properties; some of it depends on examples discussed in § 4.

(10) (a) Šmul'yan [5] proved that U is (UR) if and only if U^π is (UF); Day [4] showed that U is (UR) if and only if U^π is uniformly flattened, a condition closely equivalent to (US), which is described here about as in Bourbaki [2], Chap. V, 1, (15), but is more awkward to handle. This duality is complete for either property implies reflexivity. [Day [4] reduces the problem to two dimensions by means of two lemmas: N is (UR) [(US)] if and only if the two-dimensional *subspaces* of N have a common modulus of convexity $\delta(\varepsilon)$, [smoothness

$\varepsilon(\eta)$], (obvious) and N is (UR) [(US)] if and only if the two-dimensional *factor spaces* of N have a common modulus of convexity [smoothness]. Then formulas to calculate $\delta(\varepsilon)$ in B^* from $\varepsilon(\eta)$ in B and conversely in the two-dimensional case complete the proof.] (b) Šmul'yan [7] proved that U^π is (F) if and only if U is (ν); since (ν) implies (ρ), this gives a sketch of Šmul'yan' *s* proof that if B^* is (F), then B is reflexive. Šmul'yan also proved a dual condition that U is (F) if and only if U^π satisfies a condition (ν^*) where the functionals used must come from QB, not generally in B^{**}. (c) Šmul'yan [6] also showed that B^* is (UR$^{Q\Sigma}$) if and only if B is (UG), and a dual result. (d) Lovaglia proved that if N^* is (LUR), then N is (F). The converse need not hold since the dual of (F) is strong exposure of every support functional in Σ; see the addendum to this section on Fenchel duality for one style of proof of this.

(11) Inheritance of these properties. (a) Subspaces of spaces with one of these properties inherit the property. (b) Factor spaces usually do not; (UR) and (US) are a helpful exception. Substitution spaces Defn. II, 2, 2 behave about as well as can be expected, since copies of X and of all N_s can be found in $P_X N_s$. More precisely (c) $P_X N_s$ is (UR) if and only if all of the spaces X and N_s have a common modulus of rotundity (Day [3]). (d) Lovaglia shows that $P_X N_s$ is (LUR) if and only if X and N_s are all (LUR). (e) If $P_X N_s$ is directionally uniformly rotund, then X and all N_s must be. (f) The converse is not known (it is improbable), but if all N_s are (UR$_\Sigma$) and if X is (UR$^{A'}$) where A' is the set of evaluation functionals on X, then X and $P_X N_s$ are (UR$_\Sigma$). If X and all N_s are (UR$^{\Sigma'}$), so is $P_X N_s$. (g) If A' is a fundamental set in B^*, then N is (UR$^{A'}$) if and only if it is (UR$^{\Sigma'}$).

(12) As we observed after III, 4, (O), Dixmier [1] showed that the fourth conjugate of B can not be rotund if B is not reflexive. The example of James [2] described in IV, § 3, has a separable fourth conjugate which, by Th. 4, 1(a), is *isomorphic* to a locally uniformly rotund space B'; nevertheless, B' can not be *isometric* with any fourth conjugate space.

(13) Clarkson estimated the functions $\delta_p(\varepsilon)$ for the spaces l^p and L^p; Hanner tightened the estimate for $p<2$. If $1<p\le 2$, then

$$\left(1-\delta_p(\varepsilon)+\frac{\varepsilon}{2}\right)^p+\left(1-\delta_p(\varepsilon)-\frac{\varepsilon}{2}\right)^p=2\,;$$

if $2\le p<\infty$, then

$$\delta_p(\varepsilon)=1-\left(1-\left(\frac{\varepsilon}{2}\right)^p\right)^{\frac{1}{p}}.$$

Therefore for small $\varepsilon > 0$ we have

$$\delta_p(\varepsilon) = (p-1)\frac{\left(\frac{\varepsilon}{2}\right)^2}{2} + \cdots \quad \text{if } 1 < p \leq 2;$$

$$\delta_p(\varepsilon) = \frac{\left(\frac{\varepsilon}{2}\right)^p}{p} + \cdots \quad \text{if } 2 < p < \infty.$$

(14) V.D. Milman [11] discusses many sorts of moduli, some, like those mentioned here, depending on many two-dimensional calculations [11, Chap. 1] and others basically dependent on finite codimension [11, Chap. 3], so that the vanishing of some of them is isomorphism invariant.

Fenchel duality of convex functions. (See Fenchel for finite-dimensional spaces, Brønsted or Moreau for infinite dimensional spaces, Rockafellar for applications.)

Definition 4. Let f be a real-or $+\infty$-valued function on an LCS L such that some ξ in L^* and t in R exist with $f(x) \geq \xi(x) + t$ for all x in L, and let $D(f)$ be the set of x where $f(x)$ is finite. Then the dual function f^* is the real-or $+\infty$-valued function defined in L^* by: For each ξ in L^*

$$f^*(\xi) = \sup\{\xi(x) - f(x) : x \text{ in } L\}.$$

For ϕ of the same sort on L^*, define $\phi_*(x) = \phi^*(Qx)$ for all x in L.

(15) (a) f^* is convex and w^*-lower semicontinuous, and $D(f^*)$ is convex. [For f^* is the supremum of the w^*-continuous functionals $Qx - f(x)$.] (b) ϕ_* is convex and w-lower semicontinuous and $D(\phi_*)$ is convex. (c) If $f \leq g$, then $f^* \geq g^*$, and if $\phi \leq \psi$, then $\phi_* \geq \psi_*$. (d) $f^*{}_*[\phi_*{}^*]$ is the greatest convex w- [w^*-] lower semicontinuous function dominated by $f[\phi]$. Hence $f = f^*{}_*[\phi = \phi_*{}^*]$ if and only if $f[\phi]$ is convex and w- [w^*-] lower semicontinuous. (e) If f is a convex, w-lower-semicontinuous function in a Banach space, then each core point x of $D(f)$ is an interior point of $D(f)$ and f is continuous there. [If $E_n = \{y : f(y) \leq f(x) + 1/n\}$, then E_n is closed and x is a core point of E_n so some dilation of E_n (by category theorems) contains some ball. Contracting that ball, $f(y) \leq f(x) + 1/n$ on some ball about x, so f is upper semicontinuous at x. It was already w-lower semicontinuous and therefore is n-lower semicontinuous, since these last conditions are equivalent to the corresponding closure of the supergraph of f, the convex set $G^+ f = \{(x,t) : t \geq f(x)\}$ in $B \times R$, and by Mazur's theorem $G^+ f$ is w-closed if and only if n-closed. (f) If $t > 0$, then $[(tf)^*](\xi) = t[f^*(\xi/t)]$ for all ξ in L^*. (g) If f is quadratic-homogeneous, that is,

if $f(tx)=t^2 f(x)$ for all x in L and t in R, then f^* is also quadratic-homogeneous and $(tf)^*=f^*/t$ when $t>0$. (h) If $f(x)=k\|x\|$ in N, then $f^*(\xi)=0$ for $\|\xi\|\leq 1/k$, and is $+\infty$ everywhere else. (i) If f is a constant c, then $f^*(0)=-c$, and $f^*(\xi)=+\infty$ for all other ξ in L^*. (j) If $f=\eta$, an element of L^*, then $f^*(\eta)=0$ and $f^*(\xi)=+\infty$ for all other ξ in L^*. If $f=|\eta|$, then $f^*(t\eta)=0$ for $|t|\leq 1$ and $f^*(\xi)=+\infty$ for all other ξ in L^*.

We use Asplund's version [1] of the inf-convolution rather than that of Moreau; the two are related by a simple change of variable.

Definition 5. If f and g are convex, w-[w^*-] lower semicontinuous functionals on $L[L^*]$, define the inf-convolution $f\square g$ by: For all x in L [in L^*] $(f\square g)(x)=\inf\{(f(x+y)+g(x-y))/2;\ y \text{ in } L \text{ [in } L^*]\}$.

(16) (a) With the definitions above $f\square g$ is also convex and w-[w^*-] lower semicontinuous. (b) $f\square g\leqq (f+g)/2$. (Test $y=0$ in the definition of $f\square g$.) (c) Combinations involving duals; here x is in L, ξ in L^*, λ is a real number, and c a positive number. If

$$g(x)=cf(x),\qquad f(\lambda x),\qquad f(x)+c,\qquad f(x+b),\qquad f(b-x),$$

then

$$g^*(\xi)=cf^*\left(\frac{\xi}{c}\right),\qquad f^*\left(\frac{\xi}{\lambda}\right),\qquad f^*(\xi)-c,\qquad f^*(\xi)-\xi(b),\qquad f^*(-\xi)+\xi(b).$$

(d) $(f\wedge g)^*=f^*\vee g^*$. (e) $(f\square g)^*=(f^*+g^*)/2$.
[For
$$\begin{aligned}-(f\square g)^*(\xi) &= \inf_x\{(f\square g)(x)-\xi(x)\}\\ &= \inf_x\left\{\inf_y\{(f(x+y)+g(x-y))/2-\xi(x)\}\right\}\\ &= \inf_{x,y}\{(f(x+y)-\xi(x+y))/2+(g(x-y)-\xi(x-y))/2\}\\ &= \inf_u\{(f(u)-\xi(u))/2\}+\inf_v\{(g(v)-\xi(v))/2\}\\ &= -(f^*(\xi)+g^*(\xi))/2=-[(f^*+g^*)/2](\xi).]\end{aligned}$$

(e) Dually, $[(f+g)/2]^*=f^*\square g^*$. (f) When $D(f)$ meets $D(g)$, then $(f\vee g)^*$ = greatest convex w^*-continuous h below $f^*\wedge g^*$.

(17) In the family of all convex functions γ on R which are non-negative and are continuous at 0 with $\gamma(0)=0$, let Γ_r be the set of those γ with $\gamma(t)>0$ for all $t\neq 0$, and let Γ_s be the set of all those γ with derivative zero at zero. Then $\gamma\in\Gamma_r[\Gamma_s]$ if and only if $\gamma^*\in\Gamma_s[\Gamma_r]$.

(18) When $f(x)=\|x\|^2/2$, the various rotundity conditions on the norm translate into simple conditions on f which do not have exceptional directions along rays. (a) $\|\ \|$ is (R) if and only if for all x and y in N, $f(x)-2f((x+y)/2)+f(y)>0$. (b) $\|\ \|$ is (UR) if and only if for each $\varepsilon>0$ there is $\delta(\varepsilon)>0$ such that $f(x)-2f((x+y)/2)+f(y)\geq\delta(\varepsilon)$

whenever $f(x) \leqq 1 \geqq f(y)$ and $f(x-y) \geq \varepsilon$. (c) Similar formulations can be given for ($UR_{\mathfrak{Y}}$) and (LUR). (d) f also has the same of the differentiability properties (G), (UG), (F), or (UF) that $\| \ \|$ has; for example, $\| \ \|$ is (F) if and only if for each a in N there is (unique) α in N^* and γ in Γ_s (see (17)) such that for all x in N, $f(x) \leq f(a) + \alpha(x-a) + \gamma(\|x-a\|)$. (e) Dually, if f has Fréchet derivative α at a, then $f^*(\xi) \geq f^*(\alpha) + Qa(\xi - \alpha) + \gamma^*(\|\xi - \alpha\|)$, that is, *$\alpha$ is strongly exposed (by a) on the graph of f^*.* The converse also holds; hence f is (F) if and only if at each α in N^* which supports the graph of f at some point a of N the point α is strongly exposed on f^* by a. (f) Hence if B^* is (LUR)*, then B is (F). (g) If f^* is (F)*, then f is (ν), so B is reflexive. Asplund [2] has generalized this to other convex functions. [If f^* is (F)*, then for each α in B^* there is $\mathbf{a}$ in B^{**} which is strongly exposed by α. By Cor. II, 4, 2, every thin slice off f^{**} by α has in it points of $Q(B)$, but is of small diameter by (e) raised one space. Hence $\mathbf{a}$ is in $Q(B)$, but at the same time the slice cut off f by α is non-empty and of small diameter; this is (ν).]

(19) (a) If f is continuous and convex, then for each a in the interior of $D(f)$ there is α in B^* such that $f^*(\alpha) + f(a) = \alpha(a)$. (Use Hahn-Banach theorem.) (b) α is unique if and only if f is G-differentiable at a; then α is the G-derivative. (c) f is G-differentiable to α at a if and only if the graph of $Qa + f^*(\alpha)$ meets the graph of f^* in a set containing no other points than $(\alpha, f^*(\alpha))$. Hence *f is* (G) *if f^* is* (R). (d) Dually, the graph of f^{**} meets the graph of $Q_1\alpha + f(a)$ at the single point $(Qa, f^{**}(Qa))$ if and only if f^* has G-derivative Qa at α. Hence *f is* (R) *if f^* is* (G).

§ 3. Characterizations of Inner-Product Spaces

An *inner product* (or scalar product) in a linear space E is a symmetric bilinear functional. A normed linear space E is called an *inner product space* (or generalized Euclidean space) if there is an inner product defined in E such that $\|x\|^2 = (x,x)$ for all x in E. There are many properties known for inner-product spaces which are not true for all normed spaces; many of these are strong enough restrictions to characterize inner-product spaces among normed linear space. Fréchet constructed an identity involving norms of three elements and of their sums and differences; the most useful immediate consequence was: If every *three-dimensional* linear subspace of a normed linear E is Euclidean, then E is an inner-product space. In the same Annals, P. Jordan and J. von Neumann gave the following characterization:

(JN) For every pair f, g of elements of E

$$\|f+g\|^2 + \|f-g\|^2 = 2[\|f\|^2 + \|g\|^2]:$$

(that is, the sum of the squares of the diagonals of a parallelogram equals the sum of the squares of the sides). This fact has as its major immediate consequence:

(JN_1) A normed linear space E is an inner-product space if and only if every two-dimensional subspace is Euclidean.

This fact has been used by almost all later workers in the field to simplify the sufficiency proof of a criterion.

Ficken used a condition of symmetry to get (JN):

(F) If $\|f\|=\|g\|$, then for all real a and b, $\|af+bg\|=\|bf+ag\|$; this can be restated as follows, if f, g are replaced by $f+g, f-g$:

(F′) If $\|f+g\|=\|f-g\|$, then for all real λ, $\|f+\lambda g\|=\|f-\lambda g\|$.

E. R. Lorch took several theorems of Euclidean geometry. The most memorable of his criteria (see also Aronszajn) is:

(L) The lengths of the sides of a triangle determine the lengths of the medians; that is, in terms of the norm in E, there is a non-trivial function of three real variables $F(u,v,w)$ such that for every f, g in E, $\|f+g\|=F(\|f\|,\|g\|,\|f-g\|)$.

He also gives several other relations; the first, from which he proves the sufficiency of his result above, is a weakening of (F′):

(L_1) There is a fixed real number $\gamma\neq 0, \pm 1$, such that $f, g, \in E$ and $\|f+g\|=\|f-g\|$ imply $\|f+\gamma g\|=\|f-\gamma g\|$.

This condition is also used in the proofs of his next three criteria; his fifth criterion is an inequality:

(L_5) If $\|f\|=\|g\|$, then for all real $\alpha\neq 0$, $\|\alpha f+\alpha^{-1}g\|\geqq\|f+g\|$.

Day [7] took a step in a different direction from (JN) by showing that rhombi sufficed as well as parallelograms:

(D_1) If $\|f\|=\|g\|=1$, then $\|f+g\|^2+\|f-g\|^2=4$.

The proofs in that paper are based on a geometric reformulation of (JN) which asserts that E is an inner-product space if and only if

(E) The set of points of norm one in each plane through 0 is an ellipse.

In this direction the next improvement is due to Schoenberg who showed that in the formula (D_1) the sign of equality could be replaced by either $\geqq$ or $\leqq$, the same throughout E.

(S, ~) Let $\sim$ be one of the relations $=$, $\geqq$, $\leqq$; if $\|f\|=\|g\|=1$, then

$$\|f+g\|^2+\|f-g\|^2\sim 4.$$

Schoenberg used ($S, \geqq$) to prove that *a seminormed linear space which satisfies the following Ptolmaic inequality is an inner-product space;* temporarily, we use ab for the distance between points in a metric space.

(P) If $a,b,c,d\in E$, then $ab\cdot cd+ad\cdot bc\geqq ac\cdot bd$.

A still different, but related attack is to attempt to characterize inner-product spaces in metric terms; assuming only properties statable in terms of the distance function, ab, of a metric space M, find conditions sufficient that M be isometric to an inner-product space. Refer to L.M. Blumenthal [1] for definitions; basically, it is assumed that M is a *complete, metrically convex, externally convex, metric space* provided with an embeddability property for certain subsets.

W.A. Wilson used the Euclidean four-point property:

(e4pp-0) Any set p,q,r,s of four points of M is isometrically embeddable in a Euclidean three-space.

Blumenthal in the work above and in [2] used the *weak* and *feeble* Euclidean four-point properties:

(e4pp-1) If in addition it is required that $pq+qr=pr$, the conclusion holds.

(e4pp-2) If in addition also $pq=qr$, the conclusion holds.

Day [8] observed that a still weaker condition, the *queasy* e4pp, is sufficient:

(e4pp-3) If $p,r\in M$, there is a $q\neq p$, r in M such that $pq+qr=pr$ and such that for all s in M, the set p,q,r,s is isometrically embeddable in a Euclidean space (2-space will do, of course).

Roughly speaking the sufficiency proofs of most of these metric criteria for a normed linear space to be an inner-product space fall into two classes, those which depend on (JN) through (E) and those which depend on (JN) through (F). We begin with the proof of the basic criterion.

Proof of (JN). If (f,g) exists, it is easily seen that by taking sum or difference of $\|f+g\|^2$ and $\|f-g\|^2$ one gets either (JN) or

(a) $$4(f,g)=\|f+g\|^2-\|f-g\|^2.$$

Hence (JN) is necessary and (a) determines the inner product from the norm in E. If (JN) holds, replace f by $f\pm h$ and subtract to get

$$\|f+h+g\|^2+\|f+h-g\|^2-\|f-h+g\|^2-\|f-h-g\|^2$$
$$=2[\|f+h\|^2+\|g\|^2-\|f-h\|^2-\|g\|^2].$$

or

$$4(f+g,h)+4(f-g,h)=8(f,h).$$

Divide by 4 and set $g=f$ to get $(2f,h)=2(f,h)$ (because $(0,h)=\|h\|^2-\|-h\|^2=0$). Substitute $f+g=p$, $f-g=q$ to get $(p,h)+(q,h)=2(f,h)=(2f,h)=(p+q,h)$. This is additivity of the inner product in the first variable; symmetry is obvious, so (,) is also additive in the second variable. (a) gives $(f,f)=\|f\|^2$. Hence (JN) implies that E is an inner-product space.

Most of the criteria dependent on (JN) through (E) are special cases of the general criterion due to Day [8] which is hardly memorable but which follows easily in an inner-product space upon eliminating (f,g) between the expansions of $\|\lambda f+(1-\lambda)g\|^2$ and $\|\mu f-(1-\mu)g\|^2$:

$$\text{(b)}\quad \begin{cases} \text{For all } f,g \text{ in } E \text{ and all real } \lambda,\mu \\ \mu(1-\mu)\,\|\lambda f+(1-\lambda)g\|^2+\lambda(1-\lambda)\,\|\mu f-(1-\mu)g\|^2 \\ =[\lambda+\mu-2\lambda\mu][\lambda\mu\|f\|^2+(1-\lambda)(1-\mu)\,\|g\|^2]. \end{cases}$$

Letting $\sim$ be one of the relations $=$, $\geqq$, or $\leqq$, the weakest useful consequence of this is

$$(\mathrm{D},\sim)\quad \begin{cases} \text{If } f,g\in E \text{ and } \|f\|=\|g\|=1, \text{ then there exist } \lambda \text{ and } \mu \text{ with} \\ 0<\lambda<1 \text{ and } 0<\mu<1 \text{ such that } \mu(1-\mu)\,\|\lambda f+(1-\lambda)g\|^2 \\ +\lambda(1-\lambda)\,\|\mu f-(1-\mu)g\|^2\sim[\lambda+\mu-2\lambda\mu][\lambda\mu+(1-\lambda)(1-\mu)]. \end{cases}$$

Any one of the conditions $(D,\sim)$ *implies that* E *is an inner-product space.*

Let P be a plane (i.e., a 2-dimensional linear subspace) of E and let C be the set of points of norm 1 in P. By (E) we need only prove C an ellipse. To prove $(D,\geqq)$ sufficient let C' be the ellipse of maximum area inside C. We wish to show that C and C' coincide. Let A be the set of common points of C and C'; then A is closed.

(1) A contains at least two pairs of points $\pm f$, $\pm g$.

By an affine transformation we can assume that C' is the circle $x^2+y^2=1$ and that it touches C at the points $(\pm 1,0)$. In the family of ellipses $x^2/a^2+y^2/b^2=1$ which pass through the points on C' on the lines $x=\pm y, C'$ has minimal area. If C' touched C only at $(\pm 1,0)$, one of these ellipses with a just less than 1 would be inside C; since this is impossible, by maximality of the area of C', C' also touches C at some points $\pm g\neq\pm f$. (Indeed, g is in the cone $y\geqq|x|$; see F. Behrend.)

Now let $|\cdot\cdot|$ be the norm in P for which C' is the unit sphere. Because C' is inside C, $\|h\|\leqq|h|$ for all h in P. If A is not all of C', its complement in C' is open in C'; let f and g be points of A which are end points of an open arc of the complement of A; (1) asserts that the arc is not a semicircle. Then (b) holds for $|\cdot\cdot|$ and $(D,\geqq)$ for $\|\cdot\cdot\|$, so

$$\begin{aligned} &(\lambda+\mu-2\lambda\mu)(\lambda\mu+(1-\lambda)(1-\mu)) \\ &=\mu(1-\mu)\,|\lambda f+(1-\lambda)g|^2+\lambda(1-\lambda)\,|\mu f-(1-\mu)g|^2 \\ &\geqq\mu(1-\mu)\,\|\lambda f+(1-\lambda)g\|^2+\lambda(1-\lambda)\,\|\mu f-(1-\mu)g\|^2 \\ &\geqq(\lambda+\mu-2\lambda\mu)(\lambda\mu+(1-\lambda)(1-\mu)). \end{aligned}$$

Hence equality must hold all down the chain. Since all coefficients are positive, it follows that $|\lambda f+(1-\lambda)g|=\|\lambda f+(1-\lambda)g\|$ (and also $|\mu f-(1-\mu)g|=\|\mu f-(1-\mu)g\|$). But $\lambda f+(1-\lambda)g\neq 0$; hence $(\lambda f+(1-\lambda)g)/|\lambda f+(1-\lambda)g|$ is in A and is also in the open arc of the complement of A between f and g; this contradiction proves $A=C=C'$.

The corresponding result for $(D, \leqq)$ is proved in a dual way starting from the minimal ellipse containing C. $(S, \sim)$ is, of course, the specialization of $(D, \sim)$ in which $\lambda=\mu=1/2$. To see that (P) implies $(S, \geqq)$ it suffices to use $a=f$, $b=g$, $c=-f$, $d=-g$. A criterion of Kasahara is also a specialization of $(D, \leqq)$ setting $\mu=1/2$ and bounding λ away from 0 and 1.

On the other pattern of proof, let us show how (L_1) and (L_5) imply a new criterion (M) which easily yields (F); then we prove (F) implies (E).

As Lorch observes, it is possible in (L_1) to assume $0<\gamma<1$; for changing g to $-g$ changes γ to $-\gamma$, and interchanging f with g changes γ to γ^{-1}. In this form weaken (L_1) by allowing γ to depend on f and g

(L_6) For each f, g such that $\|f+g\|=\|f-g\|$, there is a γ, $0<\gamma<1$, such that $\|f+\gamma g\|=\|f-\gamma g\|$.

If $A=A(f,g)$ is the set of numbers γ which fit this condition, continuity shows that A is closed in $(0,1)$. (Γ) shows that A cannot have a minimal positive element, so 0 is a limit point of A. Clearly, as Lorch shows, the function $\varphi(\lambda)=\|f+\lambda g\|$ is concave upward and we have just showed that it is symmetric about 0 at a set A of values of clustering at 0. Hence the graph of $\varphi(\lambda)$ has a horizontal line of support at $\lambda=0$; that is, (L_6) implies.

(M) If $\|f+g\|=\|f-g\|$, then for all real λ, $\|f\|\leqq\|f+\lambda g\|$.

We turn now to (L_5) and show that it too implies (M). (L_5) says that in the α, β plane the convex set $\{(\alpha,\beta): \alpha\geqq 0, \alpha\beta\geqq 1\}$ has no interior point of the convex set $\{(\alpha,\beta): \|\alpha f+\beta g\|\leqq\|f+g\|\}$. By the separation theorem I, 6, 4 these sets can be separated by a line which must then support both sets at the common point $(1,1)$; hence, it must be the line $\alpha+\beta=2$. Hence the line $\alpha f+(2-\alpha)g$ supports the sphere of radius $\|f+g\|$ at $f+g$, that is, for all real α, $\|\alpha f+(2-\alpha)g\|\geqq\|f+g\|$. Substituting $f+g$, and $f-g$ for f and g gives (M) after division by 2 and the substitution $\lambda=\alpha-1$.

(M) implies (F). Take $\|f+g\|=\|f-g\|$, and set up a coordinate system with f at $(0,1)$ and g at $(1,0)$. We wish to prove (F) so we suppose, for a contradiction, that for a given $\lambda>0$, $\|f+\lambda g\|=k\|f-\lambda g\|$ with $k\neq 1$; by changing the sign of g, if necessary, we can suppose that $0<k<1$.

(M) applied to $f\pm g$ and to $g\pm f$ shows that the unit sphere has a horizontal line of support at $f/\|f\|$ and a vertical line of support at $g/\|g\|$; hence the slopes of lines of support to points in the first quadrant of the unit sphere are all non-positive. Now the point $(f+\lambda g)+k(f-\lambda g)$ is at $(\lambda(1-k), 1+k)$, so is in the first quadrant. (M) applied to $f+\lambda g$ and $k(f-\lambda g)$ asserts that the slope of one line of support at the point on the unit sphere in the direction of $(\lambda(1-k), 1+k)$ is the slope of the line from $(-\lambda k, k)$ to $(\lambda, 1)$; that is, that slope is $(-k)/\lambda(1+k)$. For this to be non-positive it is necessary that $k=1$; that is, that $\|f+\lambda g\| = \|f-\lambda g\|$.

(F) implies (E). In the plane P it is an elementary consequence of (F) that for each pair of points in C, the set of points of norm one in P, there is a unique linear isometry interchanging them. Since C can consist neither exclusively of open line segments of C nor of corners of C, C can have none of either, that is,

(2) (F) implies that C is rotund and smooth.

Now let $f \in C$; then the function $\|f+g\|$ varies continuously with g on C between the numbers 0 and 2; hence there is a g in C such that $\|f+g\| = \|f-g\|$. By (F′) the line $\{f+\mu g: \mu \text{ real}\}$ is a line of support of C at f; by (2) it is the unique tangent line to C at f. Similarly $\{g+\mu f: \mu \text{ real}\}$ is the unique tangent line to C at g; hence $\pm f$ and $\pm g$ determine each other.

Given f, g on C such that $\|f+g\| = \|f-g\|$, an affine transformation gives a rectangular coordinate system so placed in P that f is at $(0,1)$ and g at $(1,0)$. (F) and (F′) then say that C is symmetric in the ordinary Euclidean sense about the four lines $x = \pm y$, $x=0$, and $y=0$. We shall show that C is the circle of radius 1 about 0. The group of motions of C is closed and includes the four reflections above and their products, one of which is the rotation by a right angle. Hence C is determined by its arc A in the first quadrant below $x=y$. Let u be a point on A with inclination $\theta = \theta(u)$ from 0 (so $0 < \theta < \pi/4$) and let $\varphi = \varphi(u)$ be the inclination of the tangent to C at u (so $\pi/2 < \varphi < 3\pi/4$); C is a circle if and only if $\varphi(u) = \theta(u) + \pi/2$ for all u in A.

If, for a contradiction, $\varphi(u) < \theta(u) + \pi/2$, and v is the point of C in the direction $\varphi(u)$, u and v are related as were f and g at the beginning of this proof, so the tangent to C at v has the inclination $\theta(u)$. If w is the point of C in the direction $\theta(u) + \pi/2$, the rotation of C by $\pi/2$ shows that the inclination of C at w is $\varphi(u) - \pi/2 < \theta(u)$. But this is a contradiction, because the inclination of the tangent is increasing along the arc from f to w, and v is in that arc. Hence $\varphi(u) \geqq \theta(u) + \pi/2$ if $u \in A$; a similar argument shows that also $\varphi(u) \leqq \theta(u) + \pi/2$ if $u \in A$, so $\varphi(u) = \theta(u) + \pi/2$ in A, and A is an arc of a circle of radius 1. Hence C is a circle of radius 1.

The purely metric conditions shall not be discussed here but for completeness Lorch's proof that (L) implies (L_1) is presented. Take $u = f+g$ and $v = f-g$ such that $\|u\| = \|v\|$. Then

$$\begin{aligned} \|2u+v\| &= \|u+u+v\| = F(\|u\|, \|u+v\|, \|-v\|) \\ &= F(\|u+v\|, \|u\|, \|v\|) = F(\|u+v\|, \|v\|, \|u\|) \\ &= \|u+2v\|. \end{aligned}$$

Hence $\|3f+g\| = \|3f-g\|$ and (L_1) holds with $\gamma = 1/3$.

(3) The subspace projection problem, apparently related to $\mathfrak{P}_\lambda$, is quite different in its effect. (a) Murray showed that if $1 \leqq p \neq 2$, then in $l^p(\omega)$ or L^p there exists a closed linear subspace which is not the range of a continuous linear projection defined on the whole space. (b) Kakutani [3] proved that if B is at least 3-dimensional, then B is an inner-product space if and only if there is a projection of B of norm 1 onto each closed linear subspace of B. (c) If B is 1- or 2-dimensional, nothing similar can be expected, for in every normed space every line through 0 is the range of a projection of norm 1. (d) If B is isomorphic to an inner-product space, then each closed linear subspace E of B is the range of a projection with bound $\leqq M$, where M does not depend on E. For the converse see Th. 1, below.

(4) Closed linear subspaces M and N are called *complementary subspaces* in an LCS L if there is a continuous linear projection P of L onto M along N; that is, with $P^{-1}(0) = N$. (a) If M and N are complementary subspaces of L, then $M+N=L$ and $M \cap N = \{0\}$. (b) In a Banach space B let $M+N$ be dense in B and $M \cap N = \{0\}$; then [Kober] M and N are complementary subspaces of B if and only if $M+N=B$. [If complementary and $(x_n) \subseteqq M$, $(y_n) \subseteqq N$ and $x_n + y_n \to z$, then $x_n = P(x_n + y_n) \to Pz \in M$, so $y_n \to z - Pz \in N$, and $z \in M+N$. If $B = M+N$, let $T(x,y) = x+y$; then T maps $M \times N$ onto B. By Banach's interior mapping theorem II, 3, 4, T^{-1} is continuous; its first component is a projection of B on M along N.] (c) Mackey [3] showed that in a separable Banach space B every closed linear subspace M has more than one *quasi-complement;* that is, a closed linear subspace N such that $M \cap N = \{0\}$ and $M+N$ is dense in B.

The complemented-subspace theorem. Lindenstrauss and Tzafriri [1] showed that the converse of (3d) is true.

Theorem 1. *If every closed linear subspace of N has a complement in N, then N is isomorphic to an inner-product space.*

The proof uses (i) Dvoretsky's theorem IV, 1, (7b) that every infinite dimensional N mimics Hilbert space; that is, for each n in ω and each $\varepsilon > 0$ there is an isomorphism T of l_n^2 into N with $\|T\| \, \|T^{-1}\| < 1+\varepsilon$, (ii) Joichi's theorem that if H parodies N (Def. 4, 1), then N is isomorphic to an inner product space, and (iii) the theorem of Davis, Dean, and Singer that if every closed linear subspace L of N is complemented, then there is a number λ such that every finite dimensional subspace L of N has projection constant $<\lambda$; that is, there is a projection P of N onto L of norm $<\lambda$. Dvoretsky's theorem still appears to require methods outside the scope of this book but we can sketch the rest of the proof.

Proof of (ii) [Joichi]. The family $\mathscr{L}$ of finite-dimensional subspaces L of N is a commutative idempotent semigroup under vector addition and is a directed system under inclusion. By Th. V, 2, 4 there is an invariant mean μ on $m(\mathscr{L})$ which extends limit on $\mathscr{L}$. [See also Day 12.] For each L in $\mathscr{L}$ let T_L be an isomorphism of L into H such $\|x\| \leqq |T_L x| \leqq \lambda \|x\|$ for all x in L. For each x in N set $f_x(L)=0$ if $x \notin L$, $=\|T_L x\|$ if $x \in L$. Let $p(x)=\mu(f_x)$; then $\|x\| \leq p(x) \leq \lambda \|x\|$ for all x in N. Since for x and y both in L, $f_x(L)$ and $f_y(L)$ satisfy (JN), p also satisfies (JN), which implies that p is a norm derived from an inner product.

Proof of (iii) [Davis, Dean, and Singer]. Let $\lambda(E)$, the projection constant of E, be $\inf\{\|P\|: P$ is a projection on $E\}$; then $\lambda(E) \leq \dim E$. If λ is an unbounded function on $\mathscr{L}$, choose E_1 in $\mathscr{L}$ so that $\lambda(E_1) \geq 1$. Choose φ_1 to be a finite subset of Σ', the set of unit vectors in N^*, so that $\|x\| \leq 2 \sup\{\beta(x): \beta \in \varphi_1\}$ for each x in E_1 and let $Y_1 = \varphi_{1\perp}$. Then the projection P_1 of $E_1 + Y_1$ along Y_1 onto E_1 is of norm ≤ 2. In Y_1 take E_2 with $\lambda(E_2) \geq 2$, take $\varphi_2 \supseteq \varphi_1$ in Σ so that $\|x\| \leq 2\varphi_2(x)$ for all x in $E_1 + E_2$, so the projection P_2 along $Y_2 = \varphi_{2\perp}$ onto $E_1 + E_2$ is of norm ≤ 2. Continue by induction and set $E = \sum E_n = \left\{\sum_n x_n: x_n \in E_n \text{ and the sum converges}\right\}$. Then E has no bounded projection P in N, because $\|(I - P_{n-1}) P_n P\| \leq 6\|P\|$, but $(I - P_{n-1}) P_n P$ is a projection on E_n so must have norm $\geq n$.

Proof of Theorem 1 [Lindenstrauss and Tzafriri, 1]. If L is an n-dimensional subspace of an infinite-dimensional N with complemented subspaces, choose λ by (iii) and let Q be a projection of norm $< \lambda$ of N on L. By (i) there is an n-dimensional subspace C of $(I-Q)N$ with an isomorphism U of C on l_n^2 such that $\|y\| \geq \|Uy\| \geq \|y\|/2$ for all y in C. Then there is an isomorphism W of L onto l_n^2 with $\|x\| \geq \|Wx\| \geq |x|/\alpha$ and α minimal possible for such isomorphisms. To seek a bound on α independent of L, let $T = U^{-1} W/2$; this is an isomorphism of L into C with $\|x\| \geq \|Tx\| \geq \|x\|/2\alpha$ for all x in L. Take a positive number μ and let $D = \{x + \mu Tx: x \in L\}$ (like the graph of T put into $L+C$), and, again by (iii), let P be a projection on D of norm $< \lambda$. We show that $\|Tx\|$ and $\|(I-Q)Px\|$ are not small simultaneously if μ is large.

Let $V = QPT$; then $(I-Q)P = \mu T(I - \mu V)$ in L. From the choices of norms already made we have $\|(I - QP\| < (\lambda + 1)\lambda$, and for x in L

$$\|(I-Q)Px\| = \mu \|T(x - \mu Vx)\| \geq \frac{\mu \|x - \mu Vx\|}{2\alpha} \geq \frac{\mu(\|x\| - \mu \|Vx\|)}{2\alpha}$$
$$\geq \frac{\mu(\|x\| - \mu \lambda^2 \|Tx\|)}{2\alpha}.$$

Define S from L into $l_n^2 \times l_n^2$ by

$$S(x) = \left(U T x,\ \frac{U(I-Q)Px}{4\lambda^2} \right).$$

Then $\|S\| \le 1$ and for each x in L

$$\|Sx\| \ge \max\left(\|UTx\|,\ \frac{\|U(I-Q)Px\|}{4\lambda^2} \right)$$

$$\ge \max\left(\frac{\|Tx\|}{2},\ \frac{\mu(\|x\| - \mu\lambda^2 \|Tx\|)}{16\alpha\lambda^2} \right).$$

By minimality of α there is an x in L with $1/\alpha \ge \|Sx\|$. If $\mu\lambda^2\|Tx\| < \|x\|/2$, use the second estimate to get

$$\|Sx\| \ge \frac{\mu\|x\|}{32\alpha\lambda^2}.$$

If $\mu\lambda^2\|Tx\| \ge \|x\|/2$, use the first estimate to get

$$\|Sx\| \ge \frac{\|x\|}{4\mu\lambda^2}.$$

Since we do not know which case holds, we know only that $1/\alpha \ge \min(\mu/32\alpha^2, 1/4\mu\lambda^2)$, so $\alpha \le \max(32\alpha\lambda^2/\mu, 4\mu\lambda^2)$. If $\mu > 32\lambda^2$, the first case is impossible so the second holds; $\alpha \le 128\lambda^4(1+\delta)$ for all $\delta > 0$ or $\alpha \le 128\lambda^4$. This yields the hypothesis of Joichi's result (ii), so N is isomorphic to an inner-product space.

(5) Lindenstrauss and Tzafriri [1] adapt their method to prove a characterization of the isomorphism classes of "c_0 or l^p" among the spaces with unconditional bases, and of "$c_0(\Gamma)$ or L^p" among boundedly σ-complete Banach lattices.

§ 4. Isomorphisms to Improve the Norm

A. Rotundity, smoothness, and convex functions

This part of this section discusses sufficient conditions for renormability of a space, mostly to get rotundity or local uniform rotundity since this is easier than getting smoothness. This is used to extend the applicability of Asplund's theorem on differentiability of convex functions in spaces B for which B^* is rotund or locally uniformly rotund.

Section 2 discussed the rotundity or smoothness actually possessed by some standard spaces and gave some information about duality of these properties. Asplund's theorem on differentiability of convex functions, Theorem 3 below, has strong smoothness hypotheses—in fact, it is required that B^* be rotund—but has a conclusion invariant under isomorphism of the space. Hence this theorem gives a new reason, stronger than my original curiosity, for attempting to renorm a space for more rotundity or smoothness. If (X) is one of these properties, say that B *is* (X)-*able*, in symbols, B is $\langle X\rangle$ [B^* is $\langle X\rangle^*$] if B [B^*] has an isomorphic [conjugate] norm which is (X).

Theorem 1. (a) *Every Banach space B with a weakly compact fundamental set, in particular, every separable B, can be renormed so that simultaneously B is* (LUR) *and B* is* (R)*, *so, by* 2.2, *B is also* (S). (b) *Every reflexive Banach space can be renormed so that simultaneously B and B* are* (LUR) *and* (F). (c) *If* B^* *is* $\langle F\rangle^*$, *therefore if* B^{**} *is* $\langle LUR\rangle^{**}$, *then B is reflexive.* (d) *If B has a separable conjugate space or if B is a* $c_0(S)$ *for some index set S, then B can be renormed so that simultaneously* B^* *is* $(LUR)^*$ *and B is* (LUR) *and* (F). (e) *If* B^* *is* $\langle R\rangle^*$ [$\langle LUR\rangle^*$], *then B is a weak* [*strong*] *differentiability space* (Def. 1.). *If B is a weak* [*strong*] *differentiability space, then the norm in B is Gateaux* [*Fréchet*] *differentiable at every point of some dense* G_δ *subset of G. Every* (WCG) *space is a weak differentiability space; every reflexive space, every* $c_0(S)$, *and every space with separable conjugate space is a strong differentiability space.*

The proof, with excursions, occupies the rest of Part A.

Cudia [1, 2] gives a survey of renorming results through the 1950's. Some examples indicate the possibilities here. All the early results are from Clarkson and from Day [5]; $\langle UR_\Sigma\rangle$ results are from Day, James, and Swaminathan.

(1) $m(S)$. (a) $m(S)$ is $\langle R\rangle$ or $\langle UR_\Sigma\rangle$ if and only if S is countable. (b) $m(S)$ is $\langle S\rangle$ if and only if S is finite. (c) $m(S)$ is (even weakly) $\langle LUR\rangle$ if and only if S is finite (Lindenstrauss [4], Th. 5.3). (d) If μ is a measure, then $M(\mu)$ is $\langle S\rangle$ or $\langle wLUR\rangle$ if and only if it is finite dimensional, and $M(\mu)$ is $\langle R\rangle$ or $\langle UR_\Sigma\rangle$ if and only if μ is finite or σ-finite. [For the negative results embed a large $m(S)$ into $M(\mu)$; for the positive results embed $M(\mu)$ into $L^1(\mu)$ as in Lemma 1 below.]

(2) $l^1(S)$. (a) $l^1(S)$ is $\langle S\rangle$ if and only if S is countable. (b) $l^1(S)$ is always $\langle LUR\rangle$ and $\langle UR_\Sigma\rangle$. [Troyanski [1]; Asplund [3]. Day [5] shows that the natural map T of $l^2(S)$ into $c_0(S)$, used with (3d) below, proves that $l^1(S)$ is $\langle R\rangle^*$ so $c_0(S)$ is $\langle S\rangle$.] (c) When S is infinite, $l^1(S)$ is $\langle R\rangle$ but $m(S)$ ($=l^1(S)^*$) is not $\langle S\rangle$, so the dual of 2, (3a), fails to hold without additional restrictions. (d) Zisler observes that by Cor. II, 2, 2, $l^1(S)$ is $\langle WUR\rangle$ if and only if S is finite. (e) For every measure μ, the space $L^1(\mu)$ is $\langle LUR\rangle$; that is, every (AL)-space is $\langle LUR\rangle$. [The proof of Troyanski [2] is given later in this section; Day [5, 6] proved only $\langle R\rangle$.] (f) Some conjugate (AL)-spaces are $\langle R\rangle$ but not $\langle R\rangle^*$ because

the underlying space is not $\langle S\rangle$. For example, if S is infinite, $m(S)^*$ is $\langle LUR\rangle$ because it is an (AL)-space, but no conjugate norm in $m(S)^*$ is (R), because no norm in $m(S)$ is (S). (g) An $L^1(\mu)$ is $\langle S\rangle$ if and only if μ is σ-finite; that is, an (AL)-space is $\langle S\rangle$ if and only if it has a Freudenthal unit.

(3) $c_0(S)$. (a) For every S, $c_0(S)$ can be given a norm for which $c_0(S)^*$ is $(LUR)^*$ and $c_0(S)$ is (LUR) and (F). [Day [5] constructed an (R) norm in $c_0(S)$:

$$\|x\| = \sup\left\{\left(\sum_{j\le n}\left|\frac{x(s_j)}{2^j}\right|^2\right)^{\frac{1}{2}} : n\in\omega,\ s_j\in S\right\}.$$

Rainwater showed that it is actually (LUR). Asplund [3] shows that $l^1(S)=c_0(S)^*$ is $\langle LUR\rangle^*$. Asplund [1] defines an averaging process for norms, given below in Th. 2, which combines these into one norm with the good properties of both.] (b) $c_0(S)$ is $\langle UR_\Sigma\rangle$ if and only if S is countable [Day, James, and Swaminathan].

The earliest positive tools are in Clarkson and in Day [5].

Lemma 1. *If there is a one-to-one, linear, continuous T carrying B into an* (R) [(UR_Σ)], *or even* $\langle R\rangle$, *space B_1, then B is* $\langle R\rangle$ [$\langle UR_\Sigma\rangle$].

Proof. If $1\le p<\infty$, then the new norm $q(x)=(\|x\|^p+\|Tx\|^p)^{1/p}$ is (R) and is between $\|x\|$ and $(1+\|T\|)\|x\|$. For $\langle UR_\Sigma\rangle$, p must be >1 [Zisler].

Corollary 1. *If there is a one-to-one linear continuous T carrying B into some $c_0(S)$, then B is* $\langle R\rangle$; *for example, every* (WCG)*-space and every conjugate of a* (WCG)*-space is* $\langle R\rangle$. (See Thm. III, 5, 1.)

Lemma 2. *If B_0^* is $\langle R\rangle^*$ (for example if B_0 is separable or if B_0 is reflexive and $\langle S\rangle$, and if T is any continuous linear function from B_0 onto a dense subset of B, then B^* is $\langle R\rangle^*$ so B is $\langle S\rangle$; that is, B can be normed so that simultaneously B is* (S) *and B^* is* $(R)^*$.

Proof. T^* is one-to-one and is w^*-w^*-continuous. B_0^* is $(R)^*$, by 2, (3b); hence $q(f)=\|f\|_{B^*}+\|T^*f\|_{B_0^*}$ is a new, rotund, isomorphic, w^*-lower semicontinuous norm in B^*; hence it is a conjugate $(R)^*$ norm in B^*; its underlying norm in B is (S) by 2, (3a).

Corollary 2. *If B has in it a dense linear continuous image of any $L^p(\mu)$, $1<p<\infty$, or of $c_0(S)$, then B^* is $\langle R\rangle^*$ and B is* $\langle S\rangle$.

[The $L^p(\mu)$ are uniformly rotund and reflexive. $c_0(S)^*$ is $\langle R\rangle^*$ because the natural mapping of $l^1(S)=c_0(S)^*$ into $l^2(S)$ is the conjugate of the natural identity map of $l^2(S)$ into $c_0(S)$.]

Corollary 3. *If B is* (WCG), *then B^* is $\langle R\rangle^*$ and B is* $\langle S\rangle$.

By Cor. III, 2, 2, there is a w^*-w-continuous one-to-one map of B^* into $c_0(S)$. By the methods of Lemmas 1 and 2 this determines a conjugate (R)* norm in B^*.

(4) Because (UR) and (US) imply reflexivity, there is full duality between ⟨UR⟩ and ⟨US⟩. The example 2, (7) shows that ⟨UR⟩ or ⟨US⟩ is a stronger condition then reflexivity; in the present context it says that $P_2(l^{q_i})$ is ⟨UR⟩ if and only if the q_i are bounded away from one and from infinity; that is $P_2(l^{q_i})$ is ⟨UR⟩ if and only if it is already (UR), and (US), and ⟨US⟩.

In order to prove Th. 1, (a) we need the result of Troyanski [2].

Lemma 3 (Troyanski). *If B is a Banach space which has a one-to-one, linear continuous map into some $c_0(S)$ and if in B there is a transfinite sequence of bounded linear operators P_α, $\omega \le \alpha \le \mu$, where μ is the first ordinal of cardinal number dens(B), and the P_α satisfy the conditions* III, 2, (12); *that is,* (i) *P_μ is the identity,* (ii) *$\|P_\alpha\| < \infty$ for each α,* (iii) *each $(P_{\alpha+1} - P_\alpha)(B)$ is separable,* (iv) *if $\omega \le \alpha \le \beta \le \mu$, then $P_\alpha P_\beta = P_\beta P_\alpha = P_\alpha$,* (v) *if for each x in B we define $\Lambda(x,\varepsilon) = \{\alpha\colon \|P_{\alpha+1}x - P_\alpha x\| > \varepsilon(\|P_{\alpha+1}\| + \|P_\alpha\|)$, then $\Lambda(x,\varepsilon)$ is a finite set, and* (vi) *if $\Lambda(x) = \bigcup_{\varepsilon>0} \Lambda(x,\varepsilon)$ and Y_x is the closed linear hull of $P_\omega(B) + \bigcup_{\alpha\in\Lambda(x)} (P_{\alpha+1} - P_\alpha)(B)$,* then *$x \in Y_x$; then B has an isomorphic* (LUR) *norm.*

Proof. Let $\mathscr{A}_n = \{A\colon A$ is a set of not more than n ordinals $\alpha, \omega \le \alpha \le \mu$, and let $\mathscr{A} = \bigcup_{n\in\omega} \mathscr{A}_n$. For each α let $(e_i^\alpha, i\in\omega)$ be a sequence fundamental in $(P_{\alpha+1} - P_\alpha)(B)$, and let $(e_i, i\in\omega)$ be a similar sequence in $P_\omega B$. For each A in $\mathscr{A}$ and m in ω define the following functions on B: $E_m^A(x)$ = distance from x to the linear hull of

$$\{e_i, e_i^\alpha\colon i \le m, \alpha\in A\}.$$

$$t(x) = \frac{\|P_{\alpha+1}x - P_\alpha x\|}{(\|P_{\alpha+1}\| + \|P_\alpha\|)}.$$

$$F^A(x) = \sum_{\alpha\in A} t_\alpha(x).$$

$$G_n(x) = \sup\{E_n^A(x) + nF^A(x)\colon A\in\mathscr{A}_n\} \quad \text{for each } n \text{ in } \omega.$$

$$G_0(x) = \|x\|.$$

Let S be a set for which an appropriate T maps B into $c_0(S)$ and let $V = \{0, -1, -2, \ldots\} \cup \{\alpha\colon \omega \le \alpha < \mu\} \cup S$. Define Φ from B into $c_0(V)$ by: $[\Phi x](-n) = 2^{-n} G_n(x)$, $[\Phi x](\alpha) = t_\alpha(x)$, and $[\Phi x](s) = Tx(s)$ for s in S. Define a new norm in B by $\|x\|' = \|\Phi x\|$ where the norm used in $c_0(V)$ is the (LUR) norm of (3a). Take a sequence (x_k) in B with $\|x_k\|' = 1 = \|x\|'$

and $\|x_k+x\|'\to 2$; then $\|\Phi x_k-\Phi x\|_{c_0(v)}\to 0$ so $t_\alpha(x_k)\to t_\alpha(x)$ for all α, $G_n(x_k)\to G_n(x)$ for each n and $\|Tx_k-Tx\|_{c_0(S)}\to 0$. This last condition implies that if a subsequence of (x_k) converges, then it must converge to x, so the rest of the proof shows that (x_k) must be norm-totally-bounded.

Given $\varepsilon>0$, choose m in ω and A'' in A so that $E_m^{A''}(x)<\varepsilon/4$; add to A'' all those β for which $t_\beta(x)\geqq\min\{t_\alpha(x)>0:\alpha\in A''\}$ to get $A'\supseteq A''$ and $b>0$ such that $E_{m(x)}^{A'}<\varepsilon/4$ and if α is in A' but β is not, then $t_\alpha(x)-t_\beta(x)\geq b$. Let $n=\max\{m$, no. of elements in A', $(\varepsilon+3\|x\|)/3b\}$ and find A in $\mathscr{A}_n$ such that $G_n(x)-(E_n^A(x)+nF_n^A(x))<\varepsilon/3$. Then $A\supseteq A'$, for if not, A must omit some α_1 in A' and also either A has $<n$ elements or A includes an element β not in A'. In either case choose D in $\mathscr{A}_n$ so that $D=A\cup\{\alpha_1\}$ or $A\cup\{\alpha_1\}\setminus\beta$. Then

$$G_n(x)-(E_n^A(x)+nF^A(x))\geq E_n^D(x)+nF^D(x)-(E_n^A(x)+nF^A(x))$$

$$=E_n^D(x)-E_n^A(x)+n(F^D(x)-F^A(x))\geq 0-\|x\|+nb>\frac{\varepsilon}{3};$$

this contradicts the choice of A. Now as k increases $G_n(x_k)$ tends to $G_n(x)$ and $F^A(x_k)$ tends to $F^A(x)$; hence if $k>k_\varepsilon$,

$$E_n^A(x_k)<G_n(x)-nF^A(x_k)<G_n(x)-nF^A(x)+\frac{\varepsilon}{3}<E_n^A(x)+\frac{2\varepsilon}{3}<\varepsilon.$$

Hence (x_k) is bounded and is within ε of a finite dimensional subspace C_ε of B; since each ball in C_ε is norm-compact, (x_k) is covered by a finite set of balls of radius ε.

This shows that every subsequence of (x_k) contains a convergent subsequence; but we saw that such a convergent subsequence must have limit x, so (x_k) converges to x.

Corollary 4 (Troyanski [2]). *Every Banach space with a (transfinite) basis is* $\langle$LUR$\rangle$. *Every* (WCG) *space is* $\langle$LUR$\rangle$. *Every abstract L-space is* $\langle$LUR$\rangle$.

Proof. Every transfinite basis satisfies the hypotheses of Lemma 3. III, 2, (12), says that every (WCG) space also is $\langle$LUR$\rangle$. By Th. VI, 4, 2 every (AL)-space is a $\prod_{l^1(S)} L_s$, where each L_s is an (AL)-space with unit. Such a space L_s is an $L^1(\mu)$ with finite μ-measure; this is (WCG) because the set of characteristic functions of μ-measurable sets in a finite measure space is a w-compact fundamental set. But by (2) $l^1(S)$ can be renormed as a full function space with (LUR) norm, and the (LUR) product of (LUR) spaces is itself (LUR), by 2, (11 d).

Before we can complete the proof of Theorem 1 we need the method of Asplund [1] of averaging convex functions.

(5) Use the Fenchel duality of convex functions from § 2. Let f_0 and g_0 be continuous convex functions on N such that for all x in N

$$g_0(x) \le f_0(x) \le (1+C)g_0(x).$$

For each $n \ge 0$ define $f_{n+1} = (f_n + g_n)/2$ and $g_{n+1} = f_n \Box g_n (= ((f_n^* + g_n^*)/2)_*$ by 2, (2e)). Then: (a) $g_n \le f_n \le (1+2^{-n}C)g_n$ for all n. (b) Hence (g_n) and (f_n) converge uniformly on any set where g_0 is bounded to some convex continuous h such that

$$g_0 \le g_1 \le \cdots \le g_n \le \cdots \le h \le \cdots \le f_m \le \cdots \le f_1 \le f_0;$$

similarly in N^* the function h^* is a limit of (f_n^*) and of (g_n^*), and

$$f_0^* \le f_1^* \le \cdots \le f_m^* \le \cdots \le h^* \le \cdots \le g_n^* \le \cdots \le g_1^* \le g_0^*.$$

(c) If g_0 and f_0 are quadratic-homogeneous, then $g_n \le f_n \le (1+4^{-n}C)g_0$.

($g_n \le f_n$ by 2, (2b). Set $a = 1+2^{-1}C$. Then for x and y in N, $f_1 \le f_0$ and $f_1 \le a g_0$ so

$$f_0(x+y) + g_0(x-y) \ge f_1(x+y) + \frac{f_1(x-y)}{a} = f_1(x+y) + a f_1\left(\frac{(x-y)}{a}\right)$$

$$\ge (1+a) f_1\left(\frac{2x}{(1+a)}\right) = \frac{4 f_1(x)}{(1+a)}.$$

Then $g_1(x) \ge 2f_1(x)/(1+a)$ or $(1+4^{-1}C)g_1(x) \ge f_1(x)$ for all x in N. Induction proves the general case.)

Theorem 2 (Asplund [1]). *If $f_0 \ge g_0$ are two functions constructed as $\| \|^2/2$ from two isomorphic norms in N and if h is the average constructed as above for f_0 and g_0, then the norm determined by h has the same of the rotundity properties* (R), (LUR), (UR_Σ), *and* (UR) *as has f_0.*

Proof. If we set $f_n = h_n + f_0/2^n$, then by (5)

$$0 \le f_n - h \le f_n - g_n \le 4^{-n} C f_0$$

so

$$(2^{-n} - 4^{-n}C) f_0 + h_n \le h \le 2^{-n} f_0 + h_n.$$

Given x and y in N these inequalities yield, because h_n is also convex,

$$h(x) - 2h\left(\frac{(x+y)}{2}\right) + h(y)$$

$$\ge 2^{-n}\left(f_0(x) - 2f_0\left(\frac{(x+y)}{2}\right) + f_0(y) - 2^{-n}C(f_0(x) + f_0(y))\right).$$

The rest of the proof involves choosing n to match the conditions on x and y. For example, if f_0 is (UR), then for each $\varepsilon > 0$ there is for

f_0 a $\delta(\varepsilon)>0$; for $f_0(x)\le 1\ge f_0(y)$ it suffices to take $2^{-n}C<\delta(\varepsilon)/4$ to have h (UR).

We turn next to the theorem of Asplund [2] on differentiation of convex functions.

Definition 1. A Banach space B is called a *weak* [*strong*] *differentiability space* if for each convex open set E and each continuous convex function f defined on E, the set of points of Gateaux [Fréchet] differentiability of f contains a dense G_δ subset D of E.

Theorem 3. (Asplund [2]). *Let B be a space for which B^* is $\langle R\rangle^*$ [$\langle LUR\rangle^*$]; that is, B can be renormed so that the conjugate norm in B^* is* (R) [(LUR)]; *then B is a weak* [*strong*] *differentiability space.*

Proof. First we define a particular dense G_δ set D in E. Since the conclusion of the theorem is independent of renorming, we may assume that B^* is already (R) or (LUR). Let $h(x)=\|x\|^2/2$, and for each $p\ge 1$ let $h_p(x)=h(px)/p=p\|x\|^2/2$ (so that the graph of h_p is similar to that of h). Let U be the open unit ball of B.

Given f, take y in E, then take $d>0$ such that f is bounded on $y+dU$, then use only numbers $p\ge 1$ and $\lambda\geqq f(y)$ such that

$$\text{(A)}\qquad \begin{aligned} h_p(x-y)+\lambda-f(x)&\ge 0 \quad \text{for all } x \text{ in } y+dU,\\ &\ge \frac{2}{p} \quad \text{if } \|x-y\|=d. \end{aligned}$$

For each such allowed choice of y, d, p, λ, and each n in ω define $G(y,d,p,\lambda,n)=\{x\colon \|x-y\|<d \text{ and } h_p(x-y)+\lambda-f(x)<1/pn\}$; for each n in ω let G_n be the union of $G(y,d,p,\lambda,n)$ over all combinations of y,d,p, and λ which satisfy (A); let D be the intersection of the sets G_n.

Each G_n is open. Also given y and d, by enlarging p($pd^2/2>2/p+\sup\{f(x)-f(y)\colon \|x-y\|<d\}$ will be large enough), λ can then be reduced to drop the graph of $h_p(x-y)+\lambda$ down close to the graph of f at some points of $y+dU$; hence $G(y,d,p,\lambda,n)$ is an open subset of $y+dU$ which is nonempty when p is large and λ suitably small. (Note that large values of p are all that matter because, by similarity, if $0\in G(y,d,p,\lambda,n)$ and $q\ge 1$, then $0\in G(y/q,\ d/q,\ pq,\ \lambda/q,\ n)$.) This shows that each G_n is dense and open in E, so D is a dense G_δ subset of E.

Some test examples: Let α and β be elements of B^* and let g be the supremum $\alpha\vee\beta$; then d can be arbitrary, even $+\infty$. In dimension one it is clear that if h is smooth and $h(x-y)+\lambda-g(x)\ge 0$ for all x, then there is a positive δ with $h(-y)+\lambda=\delta>0=g(0)$, so $0\notin G_n$ if $n>1/\delta$. In dimension two we see that compactness of the unit ball and smoothness of h imply uniform smoothness of h so that no translation by (y,λ) of the smooth graph of h will fit closer than some positive δ to the graph of g at any point in the line L where $\alpha=\beta$. By the similarity of h_p and h, $D\cap L$ is empty.

Assume that B^* is locally uniformly rotund; then we shall show that if f is not Fréchet differentiable at 0, then $0\notin D$; since the origin may be moved to $(a,f(a))$, this disposes of the general case.

Let the graph of α support that of f at (0,0); as α is not the F-derivative of f at 0 there is $\varepsilon>0$ such that for each $\rho>0$ there is b with $\|b\|<\rho$ and $f(b)-\alpha(b)\geq\varepsilon\|b\|$. Choose β, by Hahn-Banach, so that the graph of $\beta-e$ supports the graph of f at $(b,f(b))$; then $\|\alpha-\beta\|\geq\varepsilon$. [Proof. For all $t\geq 1$, $\beta(tb)-e\geq\alpha(tb)+\varepsilon\|tb\|$; divide by t and let t grow to get $[\beta-\alpha](b)\geq\varepsilon\|b\|$.] Let $\varphi=\alpha\vee(\beta-e)$ and recall the definition of the Fenchel dual of h from Def. 2, 4: $h^*(\xi)=\sup\{\xi(x)-h(x): x\in B\}$, and dually to compute h from h^*. (Then note that $e=f^*(\beta)$.) Let $\theta=(\alpha-h^*(\alpha))\vee(\beta-h^*(\beta))$ and let Z be the set where $\alpha-h^*(\alpha)$ and $\beta-h^*(\beta)$ are equal; let $\delta=\delta(\alpha,\varepsilon)$ be that positive number which exists by the assumption that h^* is (LUR) at α. One calculation is needed to bring this to a geometric property in B.

(B) If $p\geq 1$ and $w\in Z/p$, then $h_p(w)\geq\theta_p(w)+\delta/p$.

Proof. If $z\in Z$, then

$$h(z)=\sup\{\xi(z)-h^*(\xi):\xi\in B^*\}\leq\frac{[\alpha+\beta](z)}{2}-h^*\left(\frac{(\alpha+\beta)}{2}\right)$$

$$\geq\frac{(\alpha(z)+\beta(z))}{2}-\frac{(h^*(\alpha)+h^*(\beta))}{2}+\delta$$

$$=\frac{(\alpha(z)-h^*(\alpha))}{2}+\frac{(\beta(z)-h^*(\beta))}{2}+\delta=\theta(z)+\delta.$$

(B) follows by the similarity of the graphs of h_p and θ_p to those of h and θ.

Let c be the point between 0 and b where $\alpha(c)=\beta(c)-e$. Now for any choice of y, d, p, and λ which satisfy (A) and have $\|y-c\|<d$ and $pd>\|\alpha\|\vee\|\beta\|$, we have $h_p(x-y)+\lambda\geq\varphi(x)$ for all x, so $\theta_p(x-y)+\lambda\geq\varphi(x)$ for all x. Hence $\Delta(c)=h_p(c-y)+\lambda-\varphi(c)\geq h_p(c-y)-\theta_p(c-y)+\|\beta-\alpha\|\,\|c-y-Z/p\|$. Using (B), $\Delta(c)\geq\gamma/p$, where $\gamma=(\varepsilon^2/2)\vee(\varepsilon\delta/3(\|\alpha\|+\varepsilon))$.

Since a convex, continuous f is locally Lipschitz, p can be taken large enough simultaneously for all b, and hence c, near 0. For such a p, h_p and α are continuous so $h_p(-y)+\lambda-f(0)=h_p(-y)+\lambda-\alpha(0)\geq\gamma/p$. Hence $0\notin G_n(y,d,p,\lambda,h)$ if $n\geq 1/\gamma$ and p is large. It follows that $0\notin G_n$ for large n, so $0\notin D$.

Assume that B^* is rotund; we show that if f is not G-differentiable at 0 then $0\notin D$. Move the origin as before to $(a,f(a))$ and assume α supports f at (0,0). For each closed linear subspace H of α define

$B' = B/H$, $x' = x + H$, $f'(x') = \inf\{f(x): x \in x'\}$, $h'(x') = \inf\{h(x): x \in x'\}$, and so on. Then:

(a) E' is convex and open and f' is convex and continuous on it.

(b) $h'(x') = \|x'\|^2/2$.

(c) The graph of α' supports that of f' at $(0', 0)$.

(d) If $0 \in G(y, d, p, \lambda, n)$, then y', d, p, and λ satisfy (A) for h' and f', and $f(0) = 0 = f'(0')$, so $0' \in G'(y', d, p, \lambda, n)$.

(e) If $0 \in D$, then $0' \in D'$.

But if α is not the G-derivative of f at 0, then there is also a $\beta \neq \alpha$ whose graph supports that of f at $(0,0)$; let $H = \alpha_\perp \cap \beta_\perp$. Then α' and β' both support f' at $(0', 0)$. By VII, 2, (3d), B' is smooth; because it is two-dimensional, B' is uniformly smooth, so § 2 says that $0' \notin D'$. By (E) above $0 \notin D$.

(6) It is a simple computation with flat cones, small f, and large p to see that if a is a place where f has an F-derivative, then $a \in D$, with no restriction on the nature of h.

(7) This new proof does not settle whether the condition that B^* be $\langle$LUR$\rangle$ is really necessary. For example, it is not known whether F-[G-] differentiability everywhere of the norm in B is either necessary or sufficient for B to be a strong [weak] differentiability space. It seems, however, that an f and a G-differentiable norm can be given in c_0 so that $0 \in D$ but f is not G-differentiable at 0. This, if it works out, would show that this method of proof, like Asplund's is too closely tied to the specific norm to settle the definitiveness of the theorem.

We now return to check that we have all the parts needed in the proof of Theorem 1. (a) Troyanski's Lemma 2 asserts that every (WCG) space B is $\langle$LUR$\rangle$. The Amir-Lindenstrauss Corollary 3 implies that if B is a (WCG) space, then B^* can be squeezed into a $c_0(S)$ and therefore has an (R) norm, but it is w^*-lower semicontinuous so it is a conjugate norm: therefore B^* is $\langle$R$\rangle^*$. By Asplund's averaging, Theorem 2, B has a single norm which is (LUR) and is (S) because B^* is (R)*. Every separable space has a norm compact fundamental set so is (WCG).

(b) B and B^* are both (WCG) spaces so both are $\langle$LUR$\rangle$. Asplund averaging gives one norm in B for which both B and B^* are (LUR) so, by 2, (10d), both are (F).

(c) If B^{**} is (LUR), then B^* is (F) by 2, (10d). If B^* is (F), then B is (v) which is equivalent to (K) which implies reflexivity, by 2, (10b).

(d) If B has a separable conjugate, Asplund [2] shows how to give B^* a conjugate (LUR) norm. B itself comes under (a) so there is one norm for which B is (LUR) and (F) and B^* is (LUR)*. $c_0(S)$ and $l^1(S)$ are both $\langle$LUR$\rangle$, but we must check that some (LUR) norm in $l^1(S)$ is w^*-lower semicontinuous. If one proves $l^1(S)$ is $\langle$LUR$\rangle$ by applying Troyanski's Lemma 2 to the coordinate projections, then the various functions used are w^*-lower semicontinuous sublinear functionals on $l^1(S)$ so the new norm is w^*-lower semicontinuous as well as (LUR). Again averaging gives $c_0(S)$ a norm for which both B and B^* are (LUR) and B is (S).

(e) If B is a weak [strong] differentiability space, then the convex functional $\| \ \|$ in B must have Gateaux [Fréchet] derivative everywhere in some dense G_δ subset of B. Theorem 3 has the rest of Theorem 1, (e).

B. Superreflexive spaces

The purpose of this part of this section is to describe a surprising special class of reflexive spaces, which ultimately turn out to be isomorphic to uniformly rotund spaces. Meanwhile, there is some geometry and a new comparison scheme for normed spaces.

Definition 2. A normed space N *mimics* [*parodies*] a normed space C (symbol: $N \succ C$) if for each $\varepsilon > 0$ [for some $\varepsilon > 0$] and for each finite dimensional subspace L of C there is an isomorphism T of L onto some subspace of N such that $\|T\| \, \|T^{-1}\| < 1+\varepsilon$.

(8) (a) N mimics its completion. (b) Mimicry is transitive and isometry invariant, and parody is isomorphism invariant. (See (e).) (c) c_0 mimics every N. [Given L and $\varepsilon > 0$, choose f_i, $i \leq n$, in N^* so that $\|f_i\| = 1$ and for each x in L, $\|x\| < (1+\varepsilon)\sup f_i(x)$. Define $Tx = y$ by $y_i = f_i(x)$ if $i \leq n$, $y_i = 0$ if $i > n$.] (d) If m_n is the n-dimensional space with unit cube, then $P_2 m_n$ (defined in II, 2, (11)) is reflexive and mimics c_0, so it also mimics every N. Hence *there is a separable reflexive space which mimics every normed space.* (e) If N parodies a separable space C, then there is a space C' isomorphic to C such that $N \succ C'$. [Let C_n be finite dimensional, increasing subspaces whose union is dense in C; let T_n be a λ-isomorphism of C_n into N and let $p_n(x) = \|T_n x\|$ for x in C_n; take subsequences until limits exist, and set $p(x) = \lim_{i \in \omega} p_{n_i}(x)$ if $x \in \bigcup_n C_n$. If C' is made by giving the vector space of C the norm p, then C' is λ-isomorphic to C.] (f) Dvoretsky's theorem of IV, 3, (7) says that *every infinite-dimensional normed space mimics Hilbert space.* (g) Lindenstrauss and Rosenthal showed that *each N mimics N^{**}.*

More general than the construction used in the proof of 2, (7) that n-cubes cannot be squeezed comfortably into uniformly rotund spaces, we need to consider n-trees (James [7]) and the problem of compressing them into the unit ball. The basic definition proceeds in steps in Φ, the space of all ultimately zero sequences; we use the usual basis vectors δ_i, and use $\varepsilon_i = \pm 1$. The *basic* 1-*tree* is the pair $\varepsilon_1 \delta_1$. The *basic* 2-*tree* is the basic 1-tree supplemented by the two *adjacent pairs of* 2-*twigs*, $\delta_1 + \varepsilon_2 \delta_2$, $-\delta_1 + \varepsilon_3 \delta_3$. If the basic n-tree is defined with 2^n n-twigs $(x_1, \ldots, x_{2^n})$ (arranged, say, by lexicographic order in the sequences of coefficients and using only the first 2^n-1 coordinates), then the *basic* $(n+1)$*-tree* is the basic n-tree supplemented by 2^n *adjacent pairs of* $(n+1)$*-twigs* $x_i+\varepsilon_{k_i}\delta_{k_i}$, where $k_i=i-1+2^n$. The union of all the basic n-trees is the *basic* ω*-tree.* If N is a normed space, *an* n*-tree in* N is a translation by some element of N of the image under a linear transformation T from Φ into N of the basic n-tree in Φ. *An* n-ε*-tree in* N is an n-tree in N such that $\|T\delta_i\|\geqq\varepsilon/2$ for all $i\leqq 2^{n+1}-1$; that is, all *adjacent pairs of* k*-twigs are at least* ε *apart.*

Definition 3. We are interested in the relationships between the following properties of normed spaces N:

(ρ) The completion of N is reflexive.

($\Sigma\rho$) N is *superreflexive;* that is, N does not mimic any non-reflexive space.

(γ) ε*-trees grow in* N*;* that is, for each $\varepsilon>0$ there is an $n(\varepsilon)$ in ω such that no n-ε-tree lies in U, the unit ball of N.

($N\gamma$) N is not (γ); in detail, there is $\varepsilon>0$ such that for each n in ω there is an n-ε-tree in U [This is condition P_1 of James [7].]

(Q) N is *quadrate;* that is, N mimics m_2.

(NQ) N is not (Q); in detail, there is a number a with $0<a<1$ such that if $\|x\|\leqq 1\geqq\|y\|$, then $\|x+y\|<2a$ or $\|x-y\|<2a$. [This is called *uniformly non-square* in James [6 to 9]; *inquadrate* seems shorter and adequate. See Def. 2, 2.]

(9) (γ) and ($N\gamma$) are isomorphism invariant. [Applying dilations and contractions of trees, we see that if no n-ε-tree lies in U, then no n-$k\varepsilon$-tree lies in kU, so (γ) is equivalent to: *For each* k *in* ω *and* $\varepsilon>0$ *there is* $n(k,\varepsilon)$ *so large that no* n-ε*-tree with* $n\geqq n(k,\varepsilon)$ *lies in* kU; this in turn is equivalent to: *For each* $\varepsilon>0$ *the radius of* n-ε*-trees tends to infinity as* n *increases.*]

The proofs of the next theorems are drawn from James [6 to 9] and Enflo [1]. The equivalence of the first two conditions answers a question raised in Day [1] over thirty years ago; Enflo's paper answers a question raised by James in [6]. The critical steps are James's proof that ε-trees in an inquadrate space must grow and the construction in Enflo [1] of an isomorphic (UR) norm in a space where all ε-trees grow. The proof will take most of the rest of this section.

Theorem 4. *The following conditions on a normed space* N *are equivalent:*

$\langle$UR$\rangle$ *N is isomorphic to a uniformly rotund space.*

$\langle$US$\rangle$ *N is isomorphic to a uniformly smooth space.*

$\langle$NQ$\rangle$ *N is isomorphic to an inquadrate space.*

(γ) *All ε-trees grow in N.*

($\Sigma\rho$) *N is superreflexive.*

Proof. We describe first the order of work. James [6] proved (NQ) implies (ρ); this proof in the form ($N\rho$) implies (Q) is the guide to his proof in James [7] of ($N\rho$) implies ($N\gamma$). Then James [7] introduced a condition ($N\gamma_\omega$) stronger than ($N\gamma$) and showed that it implies ($N\rho$) which implies (Q). If N is ($N\gamma$) James [7] proved that there is C such that $C \prec N$ and C is ($N\gamma_\omega$); therefore ($N\rho$); therefore (Q). This carries the proof that ($N\gamma$) implies (Q), since (Q) like ($N\gamma$) uses only the existence of certain finite sets with some metric properties, and therefore passes up from C to N if $N \succ C$. Then Enflo's paper [1] constructs a new isomorphic (UR) norm in each N with (γ).

(10) ($N\rho$) *implies* (Q). (James [6]). Suppose that B, the completion of N, is not reflexive. Then there exist θ, $0<\theta<1$, and sequences (f_j) and (z_k) in the unit balls of B and B^* (as in the Banach space part of the proof of Lemma III, 2, 1, (Ja) implies (E_r)) such that

$$f_j(z_k)=\theta \quad \text{if } j\leqq k, \qquad f_j(z_k)=0 \quad \text{if } j>k.$$

If $n\in\omega$ and if $p_1<p_2<\cdots<p_{2n}$ is a sequence of positive integers, let $S(p_1,\ldots,p_{2n})$ be the set of all x in B for which $f_j(x)=(-1)^i\theta$ if $p_{2i-1}\leqq j\leqq p_{2i}$. Thinking of all p_i as tending to infinity let

$$K_n=\liminf_{p_1}\Big(\liminf_{p_2}\Big(\ldots\Big(\liminf_{p_{2n}}\left(\inf\|x\|: x\in S(p_1,\ldots,p_n)\right)\Big)\ldots\Big)\Big).$$

For each n in ω, $S(p_1,\ldots,p_{2n})$ contains the point $\sum_{i\leqq n}(z_{p_{2i}}-z_{p_{2i-1}-1})$, so $K_n\leqq 2n$. $K_{n+1}\geqq K_n$, so for each $\delta>0$ there are an $\varepsilon>0$ and an m so large that $(K_m-\varepsilon)/(K_m+\varepsilon)>1-\delta$ and $(K_{m-1}-\varepsilon)/(K_m+\varepsilon)>1-\delta$. For this m choose $p_1<q_1<p_2<p_3<q_2<q_3<\cdots<q_{2m-2}<q_{2m-1}<p_{2m}<q_{2m}$ so large that $S(p_1,\ldots,p_{2n})$ and $S(q_1,\ldots,q_{2n})$ have elements u and v for which

(a) $\|u\|<K_m+\varepsilon$ and $\|v\|<K_m+\varepsilon$.
(b) $K_m-\varepsilon<\|z\|$ if $z\in S_1=S(q_1,p_2,q_3,p_4,\ldots,q_{2m-1},p_{2m})$, and
(c) $K_{m-1}-\varepsilon<\|z\|$ if $z\in S_2=S(p_3,q_4,p_5,\ldots,p_{2m-1},q_{2m-2})$.

Then $(u+v)/2\in S_1$ and $(u-v)/2\in S_2$. Let $x=u/(K_m+\varepsilon)$ and $y=v/(K_m+\varepsilon)$; then $\|x\|<1>\|y\|$ and $\|x\pm y\|>2(1-\delta)$. (This proof is a simplification by James of his proof from [6].)

(11) ($N\rho$) implies ($N\gamma$). (James [7].) This is a modification and generalization of the preceding proof. Instead of (p_i) and (q_i), $i\leqq 2n$, we have 2^m sequences (p_{ki}), $k\leqq 2^m$, $i\leqq 2n$, and the order of selection of these is

$$p_{11}<\cdots<p_{2^m1}<p_{12}<p_{13}<p_{22}<p_{23}<\cdots<p_{2^m2}<p_{2^m3}<\cdots;$$

that is, first the singles with second subscript 1, then the pairs with second subscripts 2, 3, then pairs with second subscripts 4, 5, and so on, ending with all the singles with $i=2n$. Then the pattern of overlapping intervals where there are values θ or $-\theta$ is the same for adjacent points as it was for the p's and q's, so adjacent pairs are about θ apart and their averages are of the same pattern with slightly smaller m; proceeding carefully by induction gives an n-θ-tree in U for any $\theta<1$. This is $(N\gamma)$.

(12) We need next the condition which James [7] calls P_1^∞ but we shall call.

$(N\gamma_\omega)$. There is an $\varepsilon>0$ for which there exists an ω-ε-tree in U.

We prove that *if C is $(N\gamma_\omega)$, then C is $(N\rho)$ and C is* (Q). [Let K be the closed convex hull of the union of all the twigs of the tree in the completion B of N. Were B reflexive, then K would be w-compact. V, 1, (9) would say that there is K_1 in K with $\operatorname{diam}(K\setminus K_1)<\varepsilon/2$; Kreĭn-Mil'man says that some twig, say x_k is in $K\setminus K_1$. But then neither of the twigs branching from x_k can be in $K\setminus K_1$, so x_k cannot be either.]

(13) *If N is $(N\gamma)$, then there is $C\prec N$ such that C is $(N\gamma_\omega)$; hence C is $(N\rho)$ and* (Q). [The definition of (N_γ) requires for each n in ω a linear mapping T_n of Φ into N which carries the basic n-tree in Φ to an n-ε-tree in U. Points α of Φ which have all coordinates rational are a countable set in Φ, so if we set $p_n(\alpha)=\|T_n\alpha\|$, the Cantor diagonal process will yield convergence of a subsequence $\lim_i p_{n_i}(\alpha)=p(\alpha)$. For all n in ω and $i<2^n$, $p_n(\delta_i)\geqq\varepsilon/2$, so $p(\delta_i)\geqq\varepsilon/2$ for all i in ω, so $p(\alpha)=0$ implies $\alpha=0$. Let C be the space obtained by norming Φ with p. In this norm the basic ω-tree is an ω-ε-tree. $N\succ C$ because in any m-dimensional subspace of C the norm is as near as desired to that in some subspace of N. By (12), C is (N_ρ) and (Q).]

(14) (a) *If $C\prec N$ and C is $(N\gamma)$ or* (Q), *then so is N.* [For $(N\gamma)$ and (Q) require only existence of finite sets with metric properties.] (b) The proof of James [7] now reaches (NQ) implies (γ); that is, *if N is inquadrate, then all ε-trees grow in N.* [(13) and (14a) say that $(N\gamma)$ implies (Q); that is, that (NQ) implies (γ).] (c) (γ) is isomorphism invariant. [All that changes is the size of $n(\varepsilon)$.]

(15) (γ) *is equivalent to* $(\Sigma\rho)$. [If N is not $(\Sigma\rho)$, then (11) and (14) imply that N is $(N\gamma)$.] If N is $(N\gamma)$, then (13) says there is a non-reflexive $C\prec N$; that is, N is not $(\Sigma\rho)$.]

With this selection from James's work completed, we now turn to Enflo's construction of a (UR) norm; this requires a description of n-ε-partitions of z.

(16) For two non-zero elements of N say that *x and y have opening* $\|x/\|x\|-y/\|y\|\,\|$. A 1-ε-*partition of* z is a pair (x_1,x_2) of points of N such that $x_1+x_2=z$, $\|x_1\|=\|x_2\|$, and x_1 and x_2 have opening at least ε. $(y_1,\ldots,y_{2n+1})$ is an $(n+1)$-ε-*partition of* z if for each $i\leqq 2^n$, (y_{2i-1},y_{2i}) is a 1-ε-partition of $x_i=y_{2i-1}+y_{2i}$ and $(x_1,\ldots,x_{2^n})$ is an n-ε-partition of z. z is defined to be the 0-ε-*partition of* z. (a) If $\|z\|=1$ and if $(x_1,\ldots,x_{2^k})$ is a k-ε-partition of z, then $(2^k x_1,\ldots,2^k x_{2^k})$ is the set of k-twigs of a

k-ε-tree. By (9b) there is an $n=n(\varepsilon)$ so large that no such k-ε-tree can lie in $2U$ if $k \geqq n$. Take a number $\delta=\delta(\varepsilon)$ with $0<\delta<\inf(\varepsilon, 2^{-n})$; hereafter in this proof we shall always assume that $\varepsilon<1/8$ and use this $n(\varepsilon)$ and this $\delta(\varepsilon)$. (b) For each non-negative integer m define $c_m = 1 + \left(\sum_{i \leqq m} 4^{-i}\right)\delta/2$, and $p(z) = \inf\left\{\sum_{i \leqq 2^m} \|x_i\|/c_m;\ 0 \leqq m \leqq n \text{ and } (x_1, \ldots, x_{2^m}) \text{ an } m\text{-}\varepsilon\text{-partition of } z\right\}$. Then the 0-$\varepsilon$-partition of z gives $p(z) \leqq (1-\delta/3)\|z\|$ and $c_m<4/3$ gives $(1-\delta)\|z\|<p(z)$. p is absolutely homogeneous and non-negative and vanishes only at 0, but p might not be subadditive. Calculations show that for $m=n$ the fraction in the definition of p is $>\|z\|$, so the inf is attained for some $m<n$.

Now if $\|x\|=\|y\|=1$ and $\|x-y\| \geqq \varepsilon$, then (x,y) is a 1-ε-partition of $z=x+y$. Choose $\gamma=\gamma(\varepsilon)$ so that $0<\gamma<4^{-n}\delta$, and take ε-partitions of x and y, $(u_1, \ldots, u_{2^j})$ and $(v_1, \ldots, v_{2^k})$ such that $\left(\sum_{i \leqq 2^j} \|u_i\|\right)/c_j < p(x)+\gamma$ and $\left(\sum_{i \leqq 2^k} \|v_i\|\right) c_k < p(y)+\gamma$, and suppose that $0 \leqq j \leqq k \leqq n-1$. By adding adjacent pairs of twigs in the tree for y, we can reduce stepwise to the j-ε-partition of y, $(w_1, \ldots, w_{2^j})$ and by the triangle rule $\sum_{i \leqq 2^j} \|w_i\|/c_k \leqq \sum_{i \leqq 2^k} \|v_i\|/c_k < p(y)+\gamma$. Pasting the (u_i) and (w_i) together into one $(k+1)$-ε-partition, of z, we can show (using the numerical values of the c_m) that if $\eta=\eta(\varepsilon)$ is about $4^{-n-2}\delta$, then $p(x+y)<p(x)+p(y)-\eta(\varepsilon)$; that is, we have the beginnings of uniform rotundity but not yet a norm.

(17) With $0<\varepsilon<1/8$ and $n(\varepsilon)$, $\delta(\varepsilon)$, and $\eta(\varepsilon)$ chosen as in (16), let $q(z)$ be the inf of the p-lengths of polygons connecting 0 to z. Then q is a norm which can be shown (as p did) to satisfy $(1-\delta)\|z\| \leqq q(z) \leqq (1-\delta/3)\|z\|$. We wish to show that q also has some uniform rotundity; more precisely, if $\|x\| = \|y\| = 1$ and $\|x-y\| \geqq 5\varepsilon$, then $q(x+y)<q(x)+q(y)-\varepsilon\eta$.

Choose $\gamma>0$ small enough that $0<\gamma+\delta<\varepsilon$ and let $0=a_0, a_1, \ldots, a_n = x$, and $0=b_0, b_1, \ldots, b_m = y$ be two chains of corners of polygons with p-lengths $<q(x)+\gamma$ and $q(y)+\gamma$, respectively. Suppose that the first chain is shorter when measured with the original norm; introduce new division points where needed in both chains and renumber, and then set $u_i = a_{i+1}-a_i$ and $v_i = b_{i+1}-b_i$ to get $\|u_i\| = \|v_i\|$ when $0 \leqq i < n$. Then $\|b_n - y\| < \varepsilon$ because the closeness of $p(u_i)$ to $\|u_i\|$ show that $\sum_{n \leqq i < m} \|v_i\| < \gamma+\delta<\varepsilon$. This means that

$$4\varepsilon \leqq \|x-b_n\| = \Big\| \sum_{0 \leqq i < n} (u_i - v_i) \Big\| \leqq \sum_{0 \leqq i < n} \|u_i - v_i\| .$$

Split this sum into those terms where the opening between u_i and v_i is less than ε, indicated by Σ', and the remaining terms, indicated by Σ''. The first sum is less than ε, so $\Sigma''\|u_i-v_i\|\geqq 3\varepsilon$, and no single term can contribute more than $2\|u_i\|$, so $\Sigma''\|u_i\|>3\varepsilon/2>\varepsilon$. By the last inequality of (16) $\Sigma'' p(u_i+v_i)\leqq\Sigma'' p(u_i)+\Sigma'' p(v_i)-\varepsilon\eta$. Gathering all the terms for x and for y together, this gives an upper estimate $q(x+y)\leqq q(x)+q(y)+2\gamma-\varepsilon\eta$. Since γ can be chosen arbitrarily small, q satisfies the desired condition.

(18) Let q_n be the function just defined in (17) for $\varepsilon=2^{-n-4}$ and set $f=\sum_{n\in\omega} q_n/2^n$. This series of norms converges to a norm which satisfies $(7/8)\|x\|\leqq f(x)\leqq\|x\|$ and is uniformly rotund; indeed, if $f(x)=1=f(y)$ and $f(x-y)\geqq\zeta$, choose n so large that $5/2^{n+4}<\zeta$; then $f(x+y)\leqq f(x)+f(y)-2^{-n-1}2^{-n-4}\eta(2^{-n-4})$.

(19) Let us now gather the bits to complete the proof of Theorem 4. (a) If B is (Q), so is B^*. [If $\|x\|=\|y\|=1$ in B and if $\|x\pm y\|\geqq a$, choose f and g supporting the unit ball at $(x+y)/\|x+y\|$ and $(x-y)/\|x-y\|$, respectively. Then $\|f\|=\|g\|=1$ and $\|f\pm g\|\geqq 2a-1$, which increases to one as a does.] (b) B is $\langle$NQ$\rangle$ if and only if B^* is $\langle$NQ$\rangle$. [If B or B^* is $\langle$NQ$\rangle$, then both are reflexive, so both are conjugate spaces; (a) applies after some isomorphism.] (c) B is $\langle$NQ$\rangle$ if and only if B is (γ) and if and only if B is $\langle$UR$\rangle$. [(13) and (14) say that (NQ) implies (γ); (14c) says (γ) is isomorphism invariant, so $\langle$NQ$\rangle$ implies (γ). (16) says that (γ) implies $\langle$UR$\rangle$.] (d) $\langle$US$\rangle$ is equivalent to $\langle$UR$\rangle$. [B is $\langle$US$\rangle$ if and only if B^* is $\langle$UR$\rangle$. By (c) this can happen if and only if B^* is $\langle$NQ$\rangle$; that is, by (a), if and only if B is $\langle$NQ$\rangle$. By (c) again, this occurs if and only if B is $\langle$UR$\rangle$. This and (15) complete the proof of equivalence.]

(20) The basic n-tree is a very sketchy part of the (2^n-1)-cube in c_0. (a) The example $P_2 m_n$ cannot be a $\langle$UR$\rangle$ space because ε-cubes can be squeezed into U for each $\varepsilon\leqq 2$. (b) Let (κ) be the condition that N parodies c_0; that is, that there is $\varepsilon>0$ such that for each n in ω some n-ε-cube can be put into U. Then (κ) implies ($N\gamma$). (c) If N parodies $c_0(\omega)$ [or $l^1(\omega)$], then N mimics $c_0(\omega)$ [or $l^1(\omega)$]. [James [6], Lemmas 2.1 and 2.2, shows that if $c_0[l^1]$ is isomorphic to a subspace of a normed space M, then $c_0[l^1]$ can be mapped arbitrarily nearly isometrically into M. (8e) asserts that if N parodies c_0 [or l^1], then it mimics an M isomorphic to c_0 [or l^1]. Hence N mimics $c_0[l^1]$. (d) Adapting the proof of (13) to (κ) instead of ($N\gamma$) says that if N has (κ), then some space C isomorphic to c_0 has $N\succ C$. By (c), $N\succ c_0$. (e) James, in a letter giving the last result (d), settles negatively a question of Day [1]; ($N\kappa$) does not imply $\langle$UR$\rangle$. James's example is l^1 which does not mimic c_0 and is not $\langle$UR$\rangle$. [To show that l^1 does not mimic m_3, show first that if $a\geqq b\geqq c\geqq 0$, the sum of the four terms $a\pm b\pm c$ is at least $5(a+b+c)/3$. Use this on each coordinate of $x\pm y\pm z$, where $x,y,z\in l^1$ and $\|x\|=\|y\|$

$= \|z\| = 1$ to get the sum of the four norms $\geqq 5$, so at least one term is $\geqq 5/4$. (f) James then asks if $(N\rho)$ is equivalent to:

(λ) N mimics l^1.

[The condition (λ) is also invariant under isomorphism, by (c).] (g) This still leaves unanswered the other problem from Day [1]: If N is not $\langle$UR$\rangle$, does N mimic $l^1(\omega)$? Using Th. 4 this is equivalent to: If N is (Q), is N also (λ)? (h) Giesy and James show that $(N\lambda)$ is equivalent to the B-convexity studied by Giesy. They prove that *if* $(N\lambda)$ *implies* (ρ), *then* $(N\lambda)$ *implies* $(\Sigma\rho)$, but they do not know if $(N\lambda)$ implies (ρ). The same paper discusses what a counter-example to this conjecture must be, and show that the space of IV, 3, Example, is not one.

(21) Akimovič shows that for most measure spaces the conditions on Φ required to make the Orlicz space $L_\Phi(\mu)$ reflexive are strong enough to make the space isomorphic to a uniformly rotund space. The exceptional finite measures seem to be the atomic ones with $\lim_i \mu(x_i)/\mu(x_{i-1}) = 0$.

Chapter VIII. Reader's Guide

A: To 1956. There are many subjects which could only be briefly mentioned or completely ignored in this book, but which have been studied in detail in research papers.

For linear algebra, which is one kind of background for normed spaces, there are such general axiomatic treatments as that in Jacobson, or such elementary books as that of Šilov on finite-dimensional linear spaces.

For convex sets and the problems arising especially in finite-dimensional spaces, where "mixed volumes" furnish tools and problems not available in general, the monographs of Bonnesen and Fenchel and of Hadwiger are good basic references. For anyone interested in infinite-dimensional convex sets, from almost any aspect, the works of V. L. Klee, some of which are listed in the bibliography, are indispensable.

The first major paper on linear topological spaces is that of von Neumann [2]. The books of Nachbin [2] and of Grothendieck [6] cover quite different kinds of material very well. The material in Bourbaki [1, 2] overlaps both of those books considerably. Nakano [2] is a general text. Two major surveys of the subject have appeared as the published versions of invited addresses. The first [Hyers] gives references to most of the early papers in the field. The second [Dieudonné 1] shows how greatly the subject changed its character as work of Mackey [1, 2], Dieudonné [4], Dieudonné and Schwartz, and a long series of papers of Köthe (see the bibliography in Dieudonné [1]) on dual spaces of sequences showed the great importance of the concept of duality of locally convex spaces.

Several books are available which discuss partially ordered linear spaces. Nakano [1] includes much of his own original work in the field. Kantorovič, Vulih, and Pinsker discuss in detail operators which are continuous in the various topologies associated with the order in a space; the book has an excellent bibliography. Birkhoff [2] has some material on vector lattices. Kreĭn and Rutman discuss many topics about monotone linear operators. Kadison is concerned with structure and with representation theorems for ordered linear spaces and Banach algebras.

Historically, the growth of the theory of normed linear spaces as a subject distinct from linear algebra or the theory of real functions can be traced through the pioneering papers of F. Riesz [2], E. Helly [1, 2], and H. Hahn. Most of the early period is summarized in Banach's book, although the references are scattered through the book in assorted footnotes, rather than collected together for easy reference. After Banach's book there were many papers about special properties of particular spaces or operators; see Studia Mathematica in the 1930's.

Since Banach's book appeared, the only books on normed linear spaces and linear operators are the book of Lyusternik and Sobolev, (available in Russian or in German translation) and the small book of Kolmogorov and Fomin. While it is true that Hilbert spaces are Banach spaces, the extra structure introduced by the inner product is so great that the problems and methods are quite different. Recent books on the theory of linear operators in Hilbert space include Ahiezer and Glazman, and the last half of Riesz and Nagy; the latter has an excellent bibliography.

Another somewhat special book is that of Hille which is basically devoted to the problems of continuity and differentiability for one- or n-parameter semigroups of linear operators. For a survey of more recent results in the field see Phillips [4].

The book of Loomis gives an introduction to the enormous range of problems connected with operator representations of locally compact groups, and with Banach algebras with involution. Gel'fand, Raĭkov, and Šilov (in Russian or in English translation) recapitulate the early work on Banach algebras with a considerable bibliography. Naĭmark [1] wrote on Banach algebras with involution and in [2] described the theory of Murray and von Neumann about the dimension and trace functions connected with factors; that is, self-adjoint algebras of operators on Hilbert space which are weakly closed and have for center only the scalar multiples of the identity operator. Dixmier [2, 3] extended much of the dimension and trace theory to W^*-algebras; that is, algebras satisfying the above conditions except that the center is not restricted. Kaplansky [1] gave an axiomatic description of a somewhat larger class of algebras, called AW^*-algebras, and showed that much of the preceding theory could be carried through for these. J. von Neumann [5] gave a decomposition theory for W^*-algebras in terms of "direct integrals" of operators and spaces. This has also been extended in papers of I. Segal and Godement. All of the decomposition theory has applications to group representations; see Mackey [4],

Mautner, and also Kaplansky [2]. Kaplansky [3] gives a survey of the state of the theory of topological rings.

A little is known about topological properties of normed spaces and of their convex subsets; see Klee [3]. Mazur [2] showed that all the $L^p(\mu)$ spaces, $p \geqq 1$, where μ is Lebesgue measure, are homeomorphic. Kadec showed [1] that $c_0(\omega)$ and $l^p(\omega)$, $p \geqq 1$, are homeomorphic and [2] that all separable, uniformly rotund spaces are homeomorphic.

Almost-periodic functions furnish another topic of interest in normed linear spaces. There are books by Bohr, who started the subject, and, twenty years later, by Maak [1] and Levitan. Eberlein [2] discusses weak-almost-periodic functions on commutative groups. Maak [2] and Jacobs [1, 2] discuss elements which are almost periodic or furtive under a bounded group of operators, and show that in Hilbert space (or any uniformly rotund space with a (UR) conjugate space) the sets of these two kinds of elements are complementary closed linear subspaces.

The theory of integration appears also in normed spaces or more general linear spaces, both as a tool and as an object of study. The two-volume work of Bourbaki [3, 4] studies integration as a process of extension of linear functionals defined on continuous function spaces. During the ten or fifteen years after Banach's book appeared Bochner, Dunford [1, 2], G. Birkhoff [1], Pettis [2], Phillips [3], and Rickart defined more and more general integrals of functions with values in a normed or a locally convex space with respect to a real-valued, countably-additive measure. Hildebrandt [1,3] and Fichtenholz-Kantorovich defined integrals with respect to finitely-addive, real-valued set functions in order to represent the linear functionals on $m(\omega)$ and $M(\mu)$, $\mu =$ Lebesgue measure. Gowurin, Day [11], and Price defined and studied integrals with operator-valued measures, as did Phillips [1], Grothendieck [5], and Bartle-Dunford-Schwartz. Dunford [2] was apparently the first to realize the advantages of treating the integral of a B-valued function x with respect to a real measure μ as an element of B^{**}.

Much of the work that has been done is concerned with generalizations of the Riesz representation theorem II, 2, (9), giving conditions on a linear operator sufficient that it be representable as some sort of integral with respect to a real- or operator-valued measure. The representation of operators as integrals can also be regarded as part of the problem of analysing the structure of the Banach algebra generated by the given operator; in particular, of determining the projections contained in it. This spectral theory has long been a successful part of the theory of Hermitian operators in Hilbert space. In general Banach spaces it is discussed in the book of Dunford and Schwartz, which also contains an exposition of the theory of normed spaces.

The mean ergodic theorem of von Neumann [3] came originally from problems of a measure-preserving flow in phase space, but was translated into an assertion about a one-parameter group of unitary operators and its averages over long intervals. Basically it is an operator version of a simple numerical phenomenon: If α is a complex number of absolute value $\leqq 1$, then $\lim_{n\in\omega} \sum_{i\leqq n} \alpha^i/n$ exists (and is 0 unless $\alpha=1$). More generally, if T is a linear operator of norm $\leqq 1$ on a reflexive Banach space B; then the limit, in the strong operator topology, of $\sum_{i\leqq n} T^i/n$ exists and is a projection of B onto the set of fixed points of T. A brief proof, which exploits the weak compactness of the unit ball in B, was discovered independently by F. Riesz [1] and by Yosida [1]. It was applied to Markoff processes by Yosida [2] and Kakutani and Yosida. Day [12] showed that each bounded abelian semigroup of operators in a reflexive space has averages converging to a projection. Eberlein [2] formulated the proper description of an ergodic semigroup of operators, and proved that every bounded abelian semigroup of operators in ergodic. Eberlein also showed the relation between ergodicity and Fejér's theorem on summability of Fourier series. Day [13] showed that every bounded representation of a semigroup S is ergodic if and only if there is an invariant mean on $m(S)$, and then [13, 14] continued the work of von Neumann [1] on the class of abstract semigroups with invariant means.

The book of Riesz and Nagy has an introduction to ergodic theory of the sort described here. A very good discussion of the relation between the various kinds of ergodic theorems, with a full bibliography, was prepared by Kakutani [4].

B: To 1972. Since more has been published in the general area now called functional analysis in the years since the first edition of this book than in all the time preceding it, this section will be forced to leave out many interesting topics.

(1) Köthe [3] and Kelley-Namioka are excellent general references for linear topological spaces. Schaefer [1] also talks about order and about tensor products. Peressini is about ordered linear spaces. Semadeni is a comprehensive recent book on continuous function spaces. Dunford and Schwartz and Yoshida [3] continue to be standard references for linear operators and normed spaces; also see Edwards.

(2) Duality theory led L. Schwartz to the invention of distributions and of spaces of distributions, as a tool for the application of linear topological spaces to differential equations and other problems arising from physics. For recent books see Horvath and also Trèves.

(3) The application of n-parameter semigroups of operators to problems in differential equations described in the earlier book of Hille, is carried forward in the expanded new edition by Hille and Phillips.

(4) Group algebras, Banach algebras, C^*-algebras, and W^*-algebras have had active study. The survey article by Bonsall points out questions in the theory of Banach algebras. Bonsall and Duncan give a good deal about numerical ranges of operators or of elements of Banach algebras, one of the useful tools in the study of Banach algebras and *-algebras. Halmos [2] gives a brief but elegant insight into some current problems in the theory of operators in Hilbert space.

(5) B.E. Johnson, in his work on derivations and cohomology of modules over locally compact groups, or over Banach algebras, which are built on conjugate Banach spaces, discovered that the group algebras of locally compact groups determine cohomologically trivial modules if and only if the group is amenable (Defn. V, 2, 2). This led Johnson to study amenable algebras, a concept which has begun to be used in Banach algebras and W^*-algebras.

(6) The theory of vector-valued and operator-valued set functions has leaned somewhat toward the finitely additive measures and their applications to operator theory. Uhl and Moedomo have studied Radon-Nikodym theorems and martingale theorems for vector measures. The book of Dinculeanu is the largest available reference, but Dunford-Schwartz has some material on vector measures.

(7) Lindenstrauss and Tzafriri [2, 3] have considered Orlicz sequence spaces and have shown that every such space has a subspace isomorphic to c_0 or some l^p. They also give conditions on f and F necessary and sufficient that the Orlicz sequence space l_f be isomorphic to a subspace of l_F, and have made some progress in describing when l_f is isomorphic to a complemented subspace of l_F.

(8) There is a large body of information on representations of points in a compact convex set by measures on the Choquet boundary, and many ramifications of this idea. A good place to begin is Phelps or perhaps V.D. Mil'man [12]. A paper giving interesting and important applications is by Choquet.

(9) Geometric ideas continue to be used to distinguish sets and spaces. For example, Whitley [2] uses "thickness" and "thinness" of the unit sphere to show that no isomorphism of the space c on c_0 can have $\|T\| \cdot \|T^{-1}\| < 2$. Schäffer [1, 2] and Nyokos and Schäffer distinguish among separable $L^1(\mu)$ spaces by the girth of the unit sphere; Schäffer and Sundaresan and also James and Schäffer show the relationship of (ρ) and $(\Sigma\rho)$ to the geometric condition that the girth of the unit sphere be greater than 4. (The girth is the infinmum of the lengths of symmetric closed curves on the surface $\|x\| = 1$.) On a larger

scale Lorentz discusses entropy of sets in a normed space and its relation to approximation theory. For a recent set of lecture notes on approximation theory see Holmes.

(10) Hewitt and Ross is a comprehensive study of abstract harmonic analysis for locally compact groups; volume 2 concentrates on the abelian case. The survey article by Williamson indicates directions in which some of this theory for groups can be extended to semigroups and in what ways the problems are different. Williamson, and Hewitt and Ross, devote some space to amenable semigroups but the most complete survey of that topic and its ramifications is Day [18], and, for the more precise results available for groups, Greenleaf.

(11) In view of the interest in comparison and classification of Banach spaces (see (12) below) and the success of the study of varieties of groups (see the book by Hanna Neumann) Diestel, Morris, and Saxon have defined a variety of locally convex spaces to be a family $\mathscr{V}$ of LCSs which includes isomorphic images, subspaces (not necessarily closed), factor spaces by closed subspaces, and topological products (as in II, 3, (11)) of spaces in $\mathscr{V}$. Applied to Banach spaces some of their results give stronger non-isomorphism results than some of those known before. However, for a theory of Banach spaces it is not clear to me that the topological product is appropriate. Interpreting the product as a substitution space over R^ω, the index space is a Fréchet space which is not a Banach space. Since R^ω is a relatively small space, at least in the sense that its conjugate is small, perhaps the appropriate index space for varieties of Banach spaces should be $l^2(\omega)$, the natural candidate for the "smallest" full function space.

(12) Classification of Banach spaces. There are several natural equivalence relations among Banach spaces—homeomorphism, bi-uniform homeomorphism, isomorphism, and isometry. We know a good deal about these relations among the elementary Banach spaces $c_0(S)$, $L^p(\mu)$, and $C(K)$, but not yet enough for a satisfactory theory. The beginnings of one can be found in Lindenstrauss [6] and the references there.

(a) Homeomorphisms. The first main result here is due to Kadec [5] who proved that all separable infinite-dimensional Banach spaces are homeomorphic to $l^1(\omega)$. This homeomorphism is now known to extend to separable Fréchet spaces; see Anderson and Bing for a proof that $l^2(\omega)$ is homeomorphic to R^ω. This result led to the conjecture that the least cardinal number of a fundamental set may be the characteristic invariant for homeomorphisms of Banach spaces; for reflexive spaces this has already been proved by Bessaga.

(b) Uniform homeomorphisms. Lindenstrauss [7] and Enflo [3,4] have shown that if $1 \leqq p_1 < p_2 < \infty$, then for no non-trivial μ and ν is there a biuniformly continuous mapping between $L^{p_1}(\mu)$ and $L^{p_2}(\nu)$. This contrasts with the biuniform continuity (see Day, [18], p. 38) of the restriction to the unit balls of the homeomorphism defined by Mazur [2] between $L^1(\mu)$ and $L^p(\mu)$, $1<p<\infty$. Lindenstrauss [5] remarks that it is not known whether biuniform homeomorphism forces isomorphism. Enflo [4] proves this when one space is Hilbert space.

(c) Isomorphism. The strongest result here is Milutin's theorem that if X and Y are uncountable compact metric spaces, then $C(X)$ and $C(Y)$ are isomorphic. (This answers in the affirmative the old question of Banach, p. 85, about the isomorphism of $C([0,1])$ and $C([0,1]\times[0,1])$.) The best available proof of Milutin's theorem is in Pelczynski [2]. Bessaga and Pelczynski have shown that for a countable compact metric space X the isomorphism classes are characterized by the ordinal ω^{α}, where α is the first ordinal for which the αth derived set of X is empty, so there are uncountably many distinct isomorphism classes of $C(X)$ spaces, X compact, metric, and countable but only one class for all other compact metric spaces.

For separable $L^p(\mu)$ spaces, $1\leqq p<\infty$, there are only two distinct classes, those of $l^p(\omega)$ and $L^p(\lambda)$, where λ is Lebesgue measure on $[0,1]$. (For $p=2$ these are not distinct.) The exponent p is an isomorphism invariant for $L^p(\mu)$ spaces, $1\leqq p\leqq\infty$, but the $L^\infty(\mu)$ spaces are not separable unless the measure is trivial. Pelczynski [1] proved that $m(\omega)$ and $L^\infty(\lambda)$ are isomorphic. Rosenthal proved that for finite (or even σ-finite) measures, $L^\infty(\mu)$ is isomorphic to $L^\infty(\nu)$ if and only if the dimension (that is, the smallest number of elements of a fundamental set) of $L^1(\mu)$ is the same as that of $L^1(\nu)$.

(d) Isometry. On the whole the classical Banach spaces are not isometric to each other. The representation theorems for conjugate spaces say that some of these spaces are isometric to the conjugates of others. The Banach-Stone theorem for continuous-function spaces says that $C(S)$ is isometric to $C(S')$ if and only if S is homeomorphic to S'. For separable $L^p(\mu)$ spaces, $1\leqq p<\infty$, the isometry classes for each $p\neq 2$ depend on the number of atoms and the presence or absence of a continuous part of the measure μ, so there are countably many different classes for each $p\neq 2$. (Lindenstrauss [6].)

(13) "Classical" Banach spaces. Since the prehistory of Banach spaces the elementary sequence and function spaces, $c_0(\omega)$, $l^p(\omega)$, $C(K)$, and $L^p(\mu)$, have been useful as examples and test spaces for conjectures. Many questions of isomorphism and isometry were raised in Banach, expecially around p. 245, some of which were answered down the years in such places as Studia Mathematica, but until recently there was nothing to pass for a theory for any non-trivial family $\mathscr{B}$ of Banach spaces. The kind of questions such a theory needs to answer are:

(i) Equivalence questions— What are the isometry and isomorphism classes in $\mathscr{B}$? (ii) Comparison questions—Which spaces are isometric or isomorphic to subspaces, or complemented subspaces, or quotient spaces, of other spaces in $\mathscr{B}$? Are there minimal spaces of some special simplicity? Do all spaces of $\mathscr{B}$ contain subspaces, or complemented subspaces, like some of these minimal spaces? (iii) Building questions— Are there methods for joining spaces together (such as by use of substitution spaces) which give all the spaces of $\mathscr{B}$ from the minimal ones, or from some larger basic subfamily of $\mathscr{B}$? (For example, the direct integral of von Neumann [5] gives such a method for constructing all W^*-algebras from the factors, the W^*-algebras with one-dimensional centers.) (iv) Duality questions—How does duality combine with any of the results which might be available for the preceding questions? Is the family ($\mathscr{B}$) closed under passage to duals or preduals?

(a) Starting from the separable classical spaces $c_0(\omega)$, $l^p(\omega)$, $L^p(\mu)$, $1 \leqq p < \infty$, and $C(K)$, K compact metric, we find a fair start on a theory in Lindenstrauss [6], but it requires a larger class of spaces to deal with preduals. For the isometric theory consider all spaces B such that B^* is an $L^p(\mu)$ space for some μ and some p with $1 \leqq p \leqq \infty$. For $1 < p \leqq \infty$ this turns out to give no new spaces (see Grothendieck [7] for the case $p = \infty$, the non-reflexive case), but yields only the $L^q(\mu)$, $1 \leqq q < \infty$. However, when B^* is an $L^1(\mu)$, then (see Lindenstrauss-Wulbert for organization and references for this) there are new spaces in B which include the $C(K)$ spaces and some other kinds, for example, the spaces $C_\sigma(S)$ of Defn. V, 4, 1, and, when S is a Choquet simplex, the space $A(S)$ of affine continuous functions on S. Lazar and Lindenstrauss prove that every separable predual of an $L^1(\mu)$ space is representable as a subspace Y of some $A(S)$ space, S a Choquet simplex, and that there is a projection of norm one of $A(S)$ on Y.

Lindenstrauss [6] notes that the class $\mathscr{B}$ obtained in this enlargement is closed under isometries, duals, preduals, and projections of norm one. A characterization theorem in the isometric theory described in Lindenstrauss [6] says that for a fixed p a Banach space B has B^* isometric to some $L^p(\mu)^*$ if and only if there is a sequence of n-dimensional subspaces E_n in B such that for each n in ω, E_n is isometric to l_n^p and $E_{n+1} \geqq E_n$, and $\bigcup_n E_n$ is dense in B.

(b) This isometric characterization for the preduals of $L^p(\mu)$ spaces motivates an analogous definition for the isomorphic theory. If $1 \leqq p \leqq \infty$, a Banach space B is called an $\mathscr{L}_{p,\lambda}$-space if for every finite-dimensional subspace C of B there is a finite-dimensional subspace D of B such that $D \supseteqq C$ and D is isomorphic with distortion less than λ to l_n^p, where

n is the dimension of D. B is an $\mathscr{L}_p$-space if there is a $\lambda \geqq 1$ for which B is an $\mathscr{L}_{p,\lambda}$-space. (Lindenstrauss-Pelczynski [1, 2].)

This family gives for the isomorphism theory a family of spaces resembling the earlier class in that their finite dimensional subspaces are uniformly l^p-like. ($l^p(\omega)$ parodies all $\mathscr{L}_{p,\lambda}$-spaces but not conversely.) Again the most interesting new spaces appear for $p=\infty$; the $\mathscr{L}_\infty$-spaces are more general than the preduals of $L^1(\mu)$ spaces appearing in the isometric theory. (See Lindenstrauss [6].) $\mathscr{L}_p$-spaces for $p=1, 2$, or ∞, can also be characterized by certain extension or lifting properties. (See Lindenstrauss [6].)

(14) There are currently four journals devoted to papers on functional analysis: Studia Mathematica; Teorija Funkciĭ, Funkcional'nyi Analiz i ih Priloženija (Harkov); Funkcional'nyi Analyz i ego Priloženija (Moscow); and Journal of Functional Analysis (New York). Due to the industry of Lindenstrauss, other students of Dvoretsky, and their colleagues, the Israel Journal of Mathematics (Jerusalem) also has a large share of papers on normed spaces. Otherwise papers in this field are scattered through the general mathematical journals.

Bibliography

Agnew, R. P.: Linear functionals satisfying prescribed conditions. Duke Math. J. **4**, 55—77 (1938).

Ahiezer, N. I., Glazman, I. M.: Teoriya lineĭnyh operatorov v gilbertovom prostranstve. Moscow-Leningrad 1950. Also in German translation as: Theorie der linearen Operatoren im Hilbert-Raum. Berlin 1954.

Ahiezer, N. I., Kreĭn, M. G.: On certain problems in the theory of moments. Harkov 1938 (Russian).

Akilov, G. P.: (1) On the extension of linear operations. Dokl. Akad. Nauk SSSR **57**, 643—646 (1947) (Russian); —(2) Necessary conditions for the extension of linear operations. Dokl. Akad. Nauk SSSR **59**, 417—418 (1948) (Russian).

Akimovič, B. A.: On uniformly convex and uniformly smooth Orlicz spaces. Teor. Funkciĭ Funkcional. Anal. i Priložen **15**, 114—120 (1972) (Russian).

Alaoglu, L.: Weak topologies in normed linear spaces. Ann. of Math. (2), **41**, 252—267 (1940).

Amir, D., Lindenstrauss, J.: The structure of weakly compact sets in Banach spaces. Ann. of Math. **88**, 35—46 (1968).

Anderson, R. D., Bing, R. H.: A complete elementary proof that Hilbert space is homeomorphic to the countably infinite product of lines. Bull. Amer. Math. Soc. **74**, 771—792 (1968).

Arens, R. F., Kelley, J. L.: Characterizations of the space of continuous functions over a compact Hausdorff space. Trans. Amer. Math. Soc. **62**, 499—508 (1947).

Aronszajn, N.: Caractérisation métrique de l'espace de Hilbert. C. R. Acad. Sci. Paris **201**, 811—813, 873—875 (1935).

Ascoli, G.: Sugli spazi lineari e le loro varieta lineari. Ann. Mat. Pura Appl. **10**, 33—81, 203—232 (1932).

Asplund, E.: (1) Averaged norms. Israel J. Math. **5**, 227—233 (1967); —(2) Fréchet differentiability of convex functions. Acta Math. **121**, 31—48 (1968); —(3) Boundedly Kreĭn-compact Banach spaces. Proc. of Functional Analysis Week at Aarhus in 1969, 46—50. Aarhus: Matem. Inst. Univ. 1969.

Asplund, E., Namioka, I.: A geometric proof of Ryll-Nardzewski's fixed point theorem. Bull. Amer. Math. Soc. **73**, 443—445 (1967).

Banach, S.: Théorie des opérations linéaires. Warsaw 1932.

Bartle, R. G., Dunford, N., Schwartz, J.: Weak compactness and vector measures. Canad. J. Math. **7**, 289—305 (1955).

Behrend, F.: Über einige Affininvarianten konvexer Bereiche. Math. Ann. **113**, 717—747 (1936).

Belluce, L. P., Kirk, W. A.: (1) Fixed point theorems for families of contraction mappings. Pacific J. Math. **18**, 213—217 (1968); —(2) Non-expansive mappings and fixed points in Banach spaces. Illinois J. Math. **11**, 474—479 (1967).

Bessaga, C.: Every reflexive space is homeomorphic to a Hilbert space. Bull. Acad. Polon. Sci. **15**, 397—399 (1967).

Bessaga, C., Pelczynski, A.: Spaces of continuous functions (IV). Studia Math. **19**, 53—62 (1960).

Birkhoff, G.: (1) Integration of functions with values in a Banach space. Trans. Amer. Math. Soc. **38**, 357—378 (1935); — (2) Lattice Theory. Amer. Math. Soc. Colloquium Publications, XXV, second edition 1948.

Bishop, E., Phelps, R.R.: (1) A proof that every Banach space is subreflexive. Bull. Amer. Math. Soc. **67**, 97—98 (1961); — (2) The support functionals of a convex set. Convexity. Proc. Symp. Pure Math., vol. VII, pp. 27—35. Amer. Math. Soc., Providence, R.I. 1963.

Blumenthal, L.M.: (1) Theory and applications of distance geometry. Oxford 1953; — (2) An extension of a theorem of Jordan and von Neumann. Pacific J. Math. **5**, 161—167 (1955).

Bochner, S.: Integration von Funktionen, deren Werte die Elemente eines Vektorraumes sind. Fund. Math. **20**, 262—276 (1933).

Bohnenblust, H.F., Karlin, S.: Geometrical properties of the unit sphere of Banach algebras. Ann. of Math. (2) **62**, 217—229 (1955).

Bohnenblust, H.F., Sobczyk, A.: Extension of functionals on complex linear spaces. Bull. Amer. Math. Soc. **44**, 91—93 (1938).

Bohr, H.: Fastperiodische Funktionen. Erg. Math. **1**, 5 (1932).

Bonnesen, T., Fenchel, W.: Theorie der konvexen Körper. Erg. Math. **1**, 4 (1934).

Bonsall, F.F.: A survey of Banach algebra theory. Bull. London Math. Soc. **2**, 257—274 (1970).

Bonsall, F.F., Duncan, J.: Numerical ranges of operators on normed spaces and of elements of normed algebras. Lecture Notes of London Math. Soc. **2** (1970).

Borsuk, K.: Drei Sätze über die *n*-dimensional euklidische Sphäre. Fund. Math. **20**, 177—190 (1933).

Bourbaki, N.: Eléments de mathématique. Actualités Sci. et Ind., Hermann et Cie., Paris. (1) Espaces vectoriels topologiques. Chapters I and II. ASI 1189 (1953); — (2) Espaces vectoriels topologiques. Chapters III—V. ASI 1229 (1955); — (3) Intégration. Chapters I—IV. ASI 1175 (1952); — (4) Intégration. Chapter V. ASI 1244 (1957).

Bourgin, D.G.: (1) Linear topological spaces. Amer. J. Math. **45**, 637—659 (1943); — (2) Some properties of Banach spaces. Amer. J. Math. **44**, 597—612 (1942).

Brodskiĭ, M.S., Mil'man, D.P.: On the center of a convex set. Dokl. Akad. Nauk SSSR **59**, 837—840 (1948).

Brønsted, A.: Conjugate convex functions in topological vector spaces. Mat.-Fys. Medd. Danske Vid. Selsk. (2) **34**, 27 pp. (1964).

Brouwer, L.E.J.: Über Abbildung von Mannigfaltigkeiten. Math. Ann. **71**, 97—115 (1911).

Browder, F.: Nonexpansive nonlinear operators in a Banach space. Proc. Nat. Acad. Sci. U.S.A. **54**, 1041—1044 (1965).

Charzyński, Z.: Sur les transformations isométriques des espaces du type (F). Studia Math. **13**, 94—121 (1953).

Choquet, G.: Deux exemples classiques de représentation integrale. Enseignement Math. (2) **15**, 63—75 (1969).

Clarkson, James A.: Uniformly convex spaces. Trans. Amer. Math. Soc. **40**, 396—414 (1936).

Collins, H.S.: Completeness and compactness in linear topological spaces. Trans. Amer. Math. Soc. **79**, 256—280 (1955).

Corson, H.H., Lindenstrauss, J.: On weakly compact subsets of Banach spaces. Proc. Amer. Math. Soc. **17**, 407—412 (1966).

Cudia, D.F.: (1) Rotundity. In: Convexity, Proc. Symp. Pure Math. vol. VII. Amer. Math. Soc. Providence, R.I. 1963; — (2) The geometry of Banach spaces: Smoothness. Trans. Amer. Math. Soc. **110**, 284—314 (1964).

Čech, E.: On bicompact spaces. Ann. of Math. (2) **38**, 823—844 (1937).

Dales, H.G.: Boundaries and peak points for Banach function algebras. Proc. London Math. Soc. (3) **22**, 121—136 (1971).

Davis, W.J., Dean, D.W., Singer, I.: Complemented subspaces and Λ systems in Banach spaces. Israel J. Math. **6**, 303—309 (1968).

Day, M.M.: (1) Reflexive Banach spaces not isomorphic to uniformly convex spaces. Bull. Amer. Math. Soc. **47**, 313—317 (1941); — (2) Some more uniformly convex spaces. Bull. Amer. Math. Soc. **47**, 504—507 (1941); — (3) Uniform convexity III. Bull. Amer. Math. Soc. **49**, 745—750 (1943); — (4) Uniform convexity in factor and conjugate spaces. Ann. of Math. (2) **45**, 375—385 (1944); — (5) Strict convexity and smoothness. Trans. Amer. Math. Soc. **78**, 516—528 (1955); — (6) Every L-space is isomorphic to a strictly convex space. Proc. Amer. Math. Soc. **8**, 415—417 (1957); — (7) Some characterizations of inner-product spaces. Trans. Amer. Math. Soc. **62**, 320—337 (1947); — (8) Some criteria of Kasahara and Blumenthal for inner-product spaces. Proc. Amer. Math. Soc. **10**, 92—100 (1959); — (9) A property of Banach spaces. Duke Math. J. **8**, 763—770 (1941); — (10) The spaces L^p with $0<p<1$. Bull. Amer. Math. Soc. **46**, 816—823 (1940); — (11) Operators in Banach spaces. Trans. Amer. Math. Soc. **51**, 583—608 (1942); — (12) Ergodic theorems for abelian semigroups. Trans. Amer. Math. Soc. **51**, 399—412 (1942); — (13) Means for the bounded functions and ergodicity of the bounded representations of semigroups. Trans. Amer. Math. Soc. **69**, 276—291 (1950); — (14) Amenable semigroups. Illinois J. Math. **1**, 509—544 (1957); — (15) On the basis problem in normed spaces. Proc. Amer. Math. Soc. **13**, 655—658 (1962); — (16) Convergence, closure, and neighborhoods. Duke Math. J. **11**, 181—199 (1944); — (17) Fixed-point theorems for compact convex sets. Illinois J. Math. **5**, 585—590; and Correction, same Jour. **8**, 713 (1961 and 1964); — (18) Semigroups and amenability. Semigroups: Proc. of a Sympos. New York, London: Academic Press 1969.

Day, M.M., James, R.C., Swaminathan, S.: Normed linear spaces that are uniformly convex in every direction. Canad. J. Math. **23**, 1051—1059 (1971).

Dean, D.W.: Projections in certain continuous function spaces $C(H)$ and subspaces of $C(H)$ isomorphic with $C(H)$. Canad. J. Math. **14**, 385—401 (1962).

Diestel, J., Morris, S.A., Saxon, S.A.: Varieties of locally convex topological vector spaces. Bull. Amer. Math. Soc. **77**, 799—803 (1971).

Dieudonné, J.: (1) Recent developments in the theory of locally convex vector spaces. Bull. Amer. Math. Soc. **59**, 495—512 (1953); — (2) Natural homomorphisms in Banach spaces. Proc. Amer. Math. Soc. **1**, 54—59 (1950); — (3) Sur un théorème de Šmulian. Arch. Math. **3**, 436—439 (1952); — (4) La dualité dans les espaces vectoriels topologiques. Ann. Sci. École Norm. Sup. **59**, 107—139 (1942); — (5) Sur la séparation des ensembles convexes dans un espace de Banach. Rev. Scient. **81**, 277—278 (1943).

Dieudonné, J., Schwartz, L.: La dualité dans les espaces (F) and (LF). Ann. Inst. Fourier (Grenoble) **1**, 61—101 (1949) (appeared 1950).

Dinculeanu, N.: Vector measures, 1st English edition. Oxford, New York: Pergamon Press 1966.

Dixmier, J.: (1) Sur un théorème de Banach. Duke Math. J. **15**, 1057—1071 (1948); — (2) Les anneaux d'opérateurs de classe finie. Ann. Sci. École Norm.

Sup. (3) **66**, 209—261 (1949); — (3) Les algèbres d'opérateurs dans l'espace de Hilbert (algèbres de von Neumann). Paris 1957; — (4) Les C^*-algèbres et leurs représentations. Paris: Gauthier-Villars 1964.

Dunford, N.: (1) Integration in general analysis. Trans. Amer. Math. Soc. **37**, 441—453 (1935); — (2) Uniformity in linear spaces. Trans. Amer. Math. Soc. **44**, 305—356 (1938).

Dunford, N., Pettis, B. J.: Linear operations on summable functions. Trans. Amer. Math. Soc. **47**, 323—392 (1940).

Dunford, N., Schwartz, J.: Linear operators, Part I. New York: Interscience, 1958.

Dvoretsky, A.: Some results on convex bodies and Banach spaces. Proc. Internat. Sympos. Linear Spaces, July 1960, pp. 123—160. Jerusalem: Hebrew University of Jerusalem 1961.

Dvoretzky, A., Rogers, C. A.: Absolute and unconditional convergence in normed linear spaces. Proc. Nat. Acad. Sci. U.S.A. **36**, 192—197 (1950).

Eberlein, W. A.: (1) Weak compactness in Banach spaces. Proc. Nat. Acad. Sci. U.S.A. **33**, 51—53 (1947); — (2) Abstract ergodic theorems and weak almost periodic functions. Trans. Amer. Math. Soc. **67**, 217—240 (1949).

Edwards, R. E.: Functional Analysis; theory and applications. New York: Holt, Rinehart, and Winston 1965.

Eidelheit, M.: Zur Theorie der konvexen Mengen in linearen normierten Räumen. Studia Math. **6**, 104—111 (1936).

Eilenberg, S.: Banach space methods in topology. Ann. of Math. **43**, 568—579 (1942).

Enflo, Per: (1) Banach spaces which can be given an equivalent uniformly convex norm. Israel J. Math. **13**, 281—288 (1972); — (2) A counterexample to the approximation problem. Acta Math. (to appear); — (3) On the nonexistence of uniform homeomorphisms between L_p-spaces. Ark. Mat. **8**, 103—105 (1969); — (4) Uniform structures and square roots in topological groups, I, II. Israel J. Math. **8**, 230—252, 253—272 (1970).

Fan, K., Glicksberg, I.: Some geometric properties of the sphere in a normed linear space. Duke Math. J. **25**, 553—568 (1958).

Fenchel, W.: On conjugate convex functions. Canad. J. Math. **1**, 73—77 (1949).

Fichtenholz, G., Kantorovich, L.: Sur les opérations dans l'espace des functions bornées. Studia Math. **5**, 69—98 (1934).

Ficken, F. A.: Note on the existence of scalar products in normed linear spaces. Ann. of Math. (2) **45**, 362—366 (1944).

Figiel, T.: Some remarks on Dvoretsky's theorem on almost spherical sections of convex bodies. Colloq. Math. **24**, 241—252 (1972).

Fréchet, M.: Sur la définition axiomatique d'une classe d'espaces vectoriels distanciés applicables vectoriellement sur l'espace de Hilbert. Ann. of Math. (2) **36**, 705—718 (1935).

Gantmaher, Vera: Über schwache totalstetige Operatoren. Mat. Sb. **7**, (49), 301—308 (1940) (Russian summary).

Gantmaher, V., Smul'yan, V.: Sur les espaces linéaires dont la sphère unitaire est faiblement compact. C. R. (Doklady) Acad. Sci. USSR **17**, 91—94 (1937).

Garkavi, A. L.: The best possible net and the best possible cross section of a set in a normed space. Izv. Akad. Nauk SSSR Ser. Mat. **26**, 87—106 (1962).

Gateaux, R.: Functions d'une infinité de variables indépendantes. Bull. Soc. Math. France **47**, 70—96 (1919).

Gelbaum, B. R.: Expansions in Banach spaces. Duke Math. J. **17**, 187—196 (1950).

Gel'fand, I.: Abstrakte Funktionen und lineare Operatoren. Rec. Math. (Matem. Sbornik) (46) **4**, 235—284 (1938).

Gel'fand, I., Raĭkov, D. A., Šilov, G. E.: Kommutativnye normirovannye kol'ca. Uspehi Mat. Nauk **1**, no. 2, (12), 48—146 (1946). Also: Commutative normed rings. Amer. Math. Soc. Translations. Ser. 2, Vol. 5, pp. 115—220.

Giesy, D. P.: On a convexity condition in normed linear spaces. Trans. Amer. Math. Soc. **125**, 114—146 (1966); Additions and corrections, same Trans. **140**, 511—512 (1969).

Giesy, D. P., James, R. C.: Uniformly non-l^1 and B-convex Banach spaces. Studia Math. (to appear 1973).

Godement, R.: Sur la théorie des représentations unitaires. Ann. of. Math. (2) **53**, 68—124 (1951).

Gohberg, I. C., Kreĭn, M. C.: Fundamental aspects of defect numbers, root numbers, and indices of linear operators. Uspehi Mat. Nauk (74) **XII**, No. 2, 43—118 (1957).

Goldstine, H. H.: Weakly complete Banach spaces. Duke Math. J. **4**, 125—131 (1938).

Goodner, D. A.: Projections in normed linear spaces. Trans. Amer. Math. Soc. **69**, 89—108 (1950).

Gowurin, M.: Über die Stieltjessche Integration abstrakter Funktionen. Fund. Math. **27**, 254—268 (1936).

Greenleaf, F. P.: Invariant means on topological groups and their applications. New York: van Nostrand Reinhold Co. 1969.

Grinblyum, M. M.: On the representation of a space of type (B) in the form of a direct sum of subspaces. Dokl. Akad. Nauk SSSR **70**, 749—752 (1950) (Russian).

Grothendieck, A.: (1) Sur la complétion du dual d'un espace vectoriel topologique. C. R. Acad. Sci. Paris **230**, 605—606 (1950); — (2) Critères de compacticité dans les espaces fonctionnels généraux. Amer. J. Math. **74**, 168—186 (1952); — (3) Produits tensoriels topologiques et espaces nucléaires. Mem. Amer. Math. Soc. **16** (1955); — (4) La théorie de Fredholm. Bull. Soc. Math. France **84**, 319—384 (1956); — (5) Sur les applications linéaires faiblement compactes d'espace du type $C(K)$. Canad. J. Math. **5**, 129—173 (1953); — (6) Espaces vectoriels topologiques. São Paulo 1954; — (7) Une caractérisation vectorielle métrique des espaces L^1. Canad. J. Math. **7**, 552—561 (1955).

Hadwiger, H.: Altes und neues über konvexe Körper. Basel 1955.

Hahn, H.: Über Folgen linearer Operatoren. Monatsh. Math. **32**, 1—88 (1922).

Halmos, P. R.: (1) Measure theory. New York 1950; — (2) Ten problems in Hilbert space. Bull. Amer. Math. Soc. **76**, 887—933 (1970).

Hanner, O.: On the uniform convexity of L^p and l^p. Ark. Mat. **3**, 239—244 (1956).

Helly, E.: (1) Über lineare Funktionaloperationen. Sitzgsber. Wien. Akad., Math.-naturwiss. Kl. IIa **121**, 265—297 (1912); — (2) Über Systeme linearer Gleichungen mit unendlichen vielen Unbekannten. Monatsh. Math. **31**, 60—91 (1921).

Hewitt, E., Ross, K. A.: Abstract harmonic analysis. Vols. I, II. Berlin-New York: Springer-Verlag 1963, 1970.

Hildebrandt, T. H.: (1) On bounded linear functional operations. Trans. Amer. Math. Soc. **36**, 868—875 (1934); — (2) On unconditional convergence in normed vector spaces. Bull. Amer. Math. Soc. **46**, 959—962 (1940); — (3) Integration in abstract spaces. Bull. Amer. Math. Soc. **59**, 111—139 (1953).

Hille, Einar: Functional analysis and semigroups. Amer. Math. Soc. Colloq. Publ. vol. 31. New York 1948.

Hille, Einar, Phillips, R. S.: Functional analysis and semigroups, revised edition. Providence, R. I. 1957.

Holmes, R.B.: A course on optimization and best approximation. Lecture Notes in Math. No. 257. Berlin-Heidelberg-New York: Springer-Verlag 1972.

Horvath, John.: Topological vector spaces and distributions. Reading, Mass.: Addison-Wesley 1966.

Husain, Taqdir.: The open mapping and closed graph theorems in topological vector spaces. Braunschweig 1965.

Hyers, D.H.: Linear topological spaces. Bull. Amer. Math. Soc. **51**, 1—21 (1945).

Jacobs, K.: (1) Ein Ergodensatz für beschränkte Gruppen im Hilbertschen Raum. Math. Ann. **128**, 340—349 (1954); — (2) Periodizitätseigenschaften beschränkter Gruppen im Hilbertschen Raum. Math. Z. **61**, 408—428 (1955).

Jacobson, N.: Lectures in abstract algebra. Volumes I, II. New York 1951 and 1953.

James, R.C.: (1) Bases and reflexivity of Banach spaces. Ann. of Math. (2) **52**, 518—527 (1950); — (2) A non-reflexive Banach space isometric with its second conjugate. Proc. Nat. Acad. Sci. U.S.A. **37**, 174—177 (1951); — (3) Reflexivity and the supremum of linear functionals. Ann. of Math. **66**, 159—169 (1957); — (4) Weakly compact sets. Trans. Amer. Math. Soc. **113**, 129—140 (1964); — (5) Weak compactness and reflexivity. Israel J. Math. **2**, 101—119 (1964); — (6) Uniformly non-square Banach spaces. Ann. of Math. **80**, 542—550 (1964); — (7) Some self-dual properties of normed linear spaces. Sympos. on infinite-dimensional topology. Ann. of Math. Studies **69**, 159—175 (1972); — (8) Super-reflexive spaces with bases. Pacific J. Math. **41**, 409—420 (1972); — (9) Superreflexive Banach spaces. Canad. J. Math. **24**, 896—904 (1972); — (10) Reflexivity and the supremum of linear functionals. Israel J. Math. **13**, 289—300 (1972); — (11) —, Schäffer, J.J.: Superreflexivity and the girth of spheres. Israel Journ. of Math. **11**, 398—404 (1972).

Jameson, G.J.O.: The weak-star closure of the unit ball in a hyperplane Proc. Edinburgh Math. Soc. **18** (Series II), 7—11 (1972).

Jerison, M.: (1) The space of bounded maps into a Banach space. Ann. of Math. (2) **52**, 309—327 (1950); — (2) Certain spaces of continuous functions. Trans. Amer. Math. Soc. **70**, 103—113 (1951).

Johnson, B.E.: Cohomology in Banach algebras. Mem. Amer. Math. Soc. **127** (1972).

Joichi, J.I.: Normed linear spaces equivalent to inner-product spaces. Proc. Amer. Math. Soc. **17**, 423—426 (1966).

Jordan, P., Neumann, J. von: On inner products in linear metric spaces. Ann. of Math. (2) **36**, 719—723 (1935).

Kadec, M.I.: (1) On homeomorphism of certain Banach spaces. Dokl. Akad. Nauk SSSR **92**, 465—468 (Russian); — (2) On topological equivalence of uniformly convex spaces. Uspehi Mat. Nauk **10**, no. 4, (66), 137—141 (1955) (Russian); — (3) Unconditional convergence of series in uniformly convex spaces. Uspehi Mat. Nauk **11**, no. 5, (71), 185—190 (1956) (Russian); — (4) Spaces isomorphic to a locally uniformly convex space. Izv. Vysš. Učebn. Zaved. Matematika (13) **6**, 51—57 (1959) and correction (25) **6**, 86—87 (1961); — (5) A proof of the topological equivalence of all separable infinite dimensional Banach spaces. Funkcional. Anal. i Priložen **1**, 53—62 (1967) (Russian).

Kadison, R.V.: A representation theory for commutative topological algebra. Mem. Amer. Math. Soc. **7** (1951).

Kakutani, S.: (1) Concrete representation of abstract (L)-spaces and the mean ergodic theorem. Ann. of Math. (2) **42**, 523—537 (1941); — (2) Concrete representation of abstract (M)-spaces. Ann. of Math. (2) **42**, 994—1024 (1941); — (3) Some characterizations of Euclidean space. Japan. J. Math. **16**, 93—97 (1939); — (4) Ergodic theory. Proc. Int. Congr. of Math. **2**, 128—142 (1950).

Kantorovič, L.V., Vulih, B.Z., Pinsker, A.G.: Funkčionalnyu analiz v poluuporyadocennyh prostranstvah. (Functional analysis in partially ordered spaces.) Moscow-Leningrad 1950.

Kaplansky, I.: (1) Projections in Banach algebras. Ann. of Math. (2) **53**, 235—249 (1951); — (2) The structure of certain operator algebras. Trans. Amer. Math. Soc. **70**, 219—255 (1951); — (3) Topological rings. Bull. Amer. Math. Soc. **54**, 809—826 (1948).

Karlin, S.: (1) Unconditional convergence in Banach spaces. Bull. Amer. Math. Soc. **54**, 148—152 (1948); — (2) Bases in Banach spaces. Duke Math. J. **15**, 971—985 (1948).

Kasahara, S.: A characterization of Hilbert space. Proc. Japan. Acad. **30**, 846—848 (1954).

Kelley, J.L.: (1) General topology. New York 1955; — (2) Banach spaces with the extension property. Trans. Amer. Math. Soc. **72**, 323—326 (1952).

Kelley, J. L., Namioka, I.: Linear topological spaces. Princeton, N. J.: van Nostrand 1963.

Klee, Jr., V.L.: (1) Convex sets in linear spaces. I and III. Duke Math. J. **18**, 443—466 (1951); **20**, 105—112 (1953); — (2) Separation properties of convex cones. Proc. Amer. Math. Soc. **6**, 313—318 (1955); — (3) Some topological properties of convex sets. Trans. Amer. Math. Soc. **78**, 30—45 (1955); — (4) Extremal structure of convex sets, I. Arch. Math. **8**, 234—240 (1957); — (5) Iteration of "lin" operation for convex sets. Math. Scand. **4**, 231—238 (1956); — (6) Convex bodies and periodic homeomorphisms in Hilbert space. Trans. Amer. Math. Soc. **74**, 10—43 (1953); — (7) Invariant metrics in groups. (Solution of a problem of Banach.) Proc. Amer. Math. Soc. **3**, 484—487 (1953); — (8) Some characterizations of reflexivity. Rev. Ci. (Lima) **52**, 15—23 (1950); — (9) Boundedness and continuity of linear functionals. Duke Math. J. **22**, 263—269 (1955).

Knaster, B., Kuratowski, C., Mazurkiewicz, S.: Ein Beweis des Fixpunktsatzes für n-dimensional Simplexe. Fund. Math. **XIV**, 132—137 (1929).

Kober, H.: A theorem on Banach spaces. Compositio Math. **7**, 135—140 (1939).

Kolmogoroff, A.: Zur Normierbarkeit eines allgemeinen topologischen linearen Raumes. Studia Math. **5**, 29—33 (1934).

Kolmogorov, A.N., Fomin, S.V.: Élementy teoriĭ funkciĭ i funkcional'nogo analiza. (Elements of function theory and of functional analysis.) Moscow 1954. Also in English translation, Rochester (New York) 1957.

Köthe, G.: (1) Bericht über neuere Entwicklungen in der Theorie der topologischen Vektorräume. Jber. Deutsch. Math.-Ver. **59**, 19—36 (1956); — (2) Dualität in der Funktionentheorie. J. Reine Angew. Math. **191**, 29—49 (1953); — (3) Topologische lineare Räume. Berlin-Göttingen-Heidelberg: Springer-Verlag 1960.

Kračkovskiĭ, S.N., Vinogradov, A.A.: On a criterion of uniform convexity in a space of type (B). Uspehi Nauk **7**, no. 3, (49), 131—134 (1952) (Russian).

Kreĭn, M.: Sur quelques questions de la géométrie des ensembles convexes situés dans un espace linéaire normé et complet. C.R. (Doklady) Acad. Sci. URSS **14**, 5—8 (1937).

Kreĭn, M.G., Krasnosel'skiĭ, M.A., Mil'man, D.P.: On the defect numbers of linear operators in a Banach space and on certain geometrical questions. Sb. Trud. Inst. Matem. Akad. Nauk Ukrain. SSR **11**, 97—112 (1948).

Kreĭn, M.G., Kreĭn, S.G.: (1) On an internal characterization of the set of all continuous functions defined on a bicompact Hausdorff space. C.R. (Doklady) Acad. Sci. URSS **27**, 427—430 (1940); — (2) Sur l'espace des fonctions continues définies sur un bicompact de Hausdorff et ses sousespaces semiordonnés. Rec. Math. (Mat. Sb.) **13**, (55), 1—37 (1942).

Kreĭn, M. G., Mil'man, D. P.: On extreme points of regular convex sets. Studia Math. **9**, 133—138 (1940).

Kreĭn, M. G., Rutman, M. A.: Lineĭnye operatory, ostavlyayuščie invariantnym konus v prostranstve Banaha. Uspehi Mat. Nauk **3**, no. 1, (23), 3—95 (1948). Also: Linear operations leaving invariant a cone in a Banach space. Amer. Math. Soc. Transl. **26**, 128 pp. (1950).

Kreĭn, M. G., Šmul'yan, V. L.: On regularly convex sets in the space conjugate to a Banach space. Ann. of Math. **41**, 556—583 (1940).

Lazar, A., Lindenstrauss, J.: Banach spaces whose duals are L^1-spaces and their representing matrices. Acta Math. **126**, 165—193 (1971).

Levitan, B. M.: Počti-periodičeskie funkciĭ. Moscow 1953.

Lindenstrauss, J.: (1) On reflexive spaces having the metric approximation property. Israel. J. Math. **3**, 199—204 (1965); — (2) On non-separable reflexive Banach spaces. Bull. Amer. Math. Soc. **72**, 967—970 (1966); — (3) On operators which attain their norm. Israel J. Math. **1**, 139—148 (1963); — (4) Weakly compact sets—their topological properties and the Banach spaces they generate. Sympos. on infinite dimensional topology. Ann. of Math. Studies **69**, 235—273 (1972); — (5) Some aspects of the theory of Banach spaces. Advances in Math. **5**, 159—180 (1970); — (6) Geometrical theory of the classical Banach spaces. Proc. Int. Congr. of Math. Nice in 1970 (to appear); — (7) On non-linear projections in Banach spaces. Michigan Math. J. **11**, 268—287 (1964).

Lindenstrauss, J., Pelczynski, A.: (1) Absolutely summing operators in L_p-spaces and their applications. Studia Math. **29**, 275—326 (1968); — (2) Contributions to the theory of the classical Banach spaces. J. Functional Analysis **8**, 225—249 (1971).

Lindenstrauss, J., Rosenthal, H. P.: The L_p-spaces. Israel J. Math. **7**, 325—349 (1969).

Lindenstrauss, J., Tzafriri, L.: (1) On the complemented subspaces problem. Israel J. Math. **9**, 263—269 (1971); — (2) On Orlicz sequence spaces I. Israel J. Math. **10**, 379—390 (1971); — (3) On Orlicz sequence spaces II. Israel J. Math. **11**, 355—379 (1972).

Lindenstrauss, J., Wulbert, D.: On the classification of the Banach spaces whose duals are L^1 spaces. J. Functional Analysis **4**, 332—349 (1969).

Lindenstrauss, J., Zippin, M.: Banach spaces with sufficiently many Boolean algebras of projections. J. Math. Anal. Appl. **25**, 309—320 (1969).

Loomis, L.: Abstract harmonic analysis. Toronto-New York-London: van Nostrand 1953.

Lorch, E. R.: On some implications which characterize Hilbert space. Ann. of Math. **49**, 523—532 (1948).

Lorentz, G. G.: Metric entropy and approximation. Bull. Amer. Math. Soc. **72**, 903—937 (1966).

Lovaglia, A. R.: Locally uniformly convex Banach spaces. Trans. Amer. Math. Soc. **78**, 225—238 (1955).

Löwig, H.: Über die Dimension linearer Räume. Studia Math. **5**, 18—23 (1934),

Lyusternik, L. A., Sobolev, V. I.: Elementy funkcionalnogo analiza. Moscow-Leningrad 1951 (Russian). Also: Elemente der Funktionalanalysis. Berlin 1955 (German translation).

Maak, W.: (1) Fastperiodische Funktionen. Berlin-Göttingen-Heidelberg 1950; — (2) Periodizitätseigenschaften unitärer Gruppen in Hilberträumen. Math. Scand. **2**, 334—344 (1954).

Mackey, G.W.: (1) On infinite-dimensional linear spaces. Trans. Amer. Math. Soc. **57**, 155—207 (1945); — (2) On convex topological linear spaces. Trans. Amer. Math. Soc. **60**, 519—537 (1946); — (3) Note on a theorem of Murray. Bull. Amer. Soc. **52**, 322—325 (1946); — (4) Functions on locally compact groups. Bull. Amer. Math. Soc. **56**, 385—412 (1950).

Maharam, Dorothy.: On homogeneous measure algebras. Proc. Nat. Acad. Sci. U.S.A. **28**, 108—111 (1942).

Mankiewicz, P.: On Lipschitz mappings between Fréchet spaces. Studia Math. **41**, 225—241 (1972).

Markov, A.A.: Quelques théorèmes sur les ensembles abéliens. C.R. (Doklady) Acad. Sci. URSS **1**, 311—313 (1936).

Markuševič, A.I.: On a basis (in the wide sense of the word) for linear spaces. Dokl. Akad. Nauk SSSR **41**, 241—244 (1943).

Marti, J.T.: Introduction to the theory of bases. Berlin-Heidelberg-New York: Springer-Verlag 1969.

Mautner, F.I.: Unitary representations of locally compact groups I and II. Ann. of Math. **51**, 1—25, 528—556 (1950).

Mazur, S.: (1) Über konvexe Mengen in linearen normierten Räumen. Studia Math. **4**, 70—84 (1933); — (2) Une remarque sur l'homéomorphie des champs fonctionnels. Studia Math. **1**, 83—85 (1930).

Mazur, S., Ulam, S.: Sur les transformations isométriques d'espaces vectoriels. C.R. Acad. Sci. Paris **194**, 946—948 (1932).

Michael, E.: (1) Some extension theorems for continuous functions. Pacific. J. Math. **3**, 789—806 (1953); — (2) Continuous selections I. Ann. of Math. (2) **63**, 361—382 (1956).

Mil'man, D.P.: (1) Characteristics of extreme points of regularly convex sets. Dokl. Akad. Nauk SSSR **57**, 119—122 (1947) (Russian); — (2) Accessible points of a functional compact set. Dokl. Akad. Nauk SSSR **59**, 1045—1048 (1948) (Russian); — (3) Dynamical systems defined by functionals and invariant measures on them. Dokl. Akad. Nauk SSSR **59**, 1397—1398 (1948) (Russian); — (4) Isometry and extremal points. Dokl. Akad. Nauk SSSR **59**, 1241—1244 (1948) (Russian); — (5) Multimetric spaces: Analysis of the invariant subsets of a multinormed bicompact space under a semigroup of non-increasing operators. Dokl. Akad. Nauk SSSR **67**, 27—30 (1949) (Russian); — (6) Extremal points and centers of convex bicompacta. Uspehi Mat. Nauk **4**, no. 5, (33), 179—181 (1949) (Russian); — (7) On the theory of rings with involution. Dokl. Akad. Nauk SSSR **76**, 349—352 (1951) (Russian); — (8) On some criteria for the regularity of spaces of the type (B). C.R. (Doklady) Acad. Sci. URSS **20**, 243—246 (1938).

Mil'man, D.P., Rutman, M.A.: On a more precise theorem about the completeness of the system of extremal points of a regularly convex set. Dokl. Akad. Nauk SSSR **60**, 25—27 (1948) (Russian).

Mil'man, V.D.: (9) A new proof of the theorem of *A*. Dvoretsky on sections of convex bodies. Funkcional. Anal. i Priložen **5**, 28–37 (1971) (Russian). Also in the complete English translation of this journal; — (10) Geometric theory of Banach spaces I: Theory of basic and minimal systems. Uspehi Mat. Nauk **25**, 113—173 (1970) (Russian), translated in: Russian Math. Surveys **25**, 111—170 (1970); — (11) Geometric theory of Banach spaces II: Geometry of the unit sphere. Uspehi Mat. Nauk **26**, 73—149 (1971) (Russian), translated in: Russian Math. Surveys **26**, 79—163 (1971); — (12) Facial characteristics of convex sets; extremal elements. Trudy Moskov. Mat. Obšč. **22**, 63—126 (1970) (Russian), translated in: Trans. Moscow Math. Soc. **22**, 79—140 (1970).

Milutin, A. A.: Isomorphism of spaces of continuous functions over compacta of the power of the continuum. Teor. Funkciĭ Funkcional. Anal. i Priložen **2**, 150—156 (1966).

Moreau, J. J.: Inf-convolution, sous-additivité, convexité des fonctions numériques. J. Math. Pures Appl. **49**, 109—154 (1970).

Munroe, M. E.: Measure and integration. Cambridge (Mass.) 1953.

Murray, F. J.: On complementary manifolds and projections in spaces L_p and l_p. Trans. Amer. Math. Soc. **41**, 138—152 (1937).

Murray, F. J., Neumann, J. von: On rings of operators. Ann. of Math. **37**, 116—229 (1936). — II. Trans. Amer. Math. Soc. **41**, 208—248 (1937). — IV. Ann. of Math. **44**, 716—808 (1943).

Myers, S. B.: Banach spaces of continuous functions. Ann. of Math. **49**, 132—140 (1948).

Nachbin, L.: (1) A theorem of Hahn-Banach type for linear transformations. Trans. Amer. Math. Soc. **68**, 28—46 (1950); — (2) Espaços vetoriais topologicos. Rio de Janeiro 1948.

Naĭmark, M. A.: (1) Kol'ca s involyucieĭ. Uspehi Mat. Nauk **3**, no. 5, (27), 52—145 (1948). Also: Rings with involution. Amer. Math. Soc. Transl. **25** (1950); — (2) Kol'ca operatorov v gil'bertovom prostranstve. Uspehi Mat. Nauk **4**, no. 4, (32), 83—147 (1949).

Nakano, H.: (1) Modulared semi-ordered linear spaces. Tokyo 1950; — (2) Topology and linear topological spaces. Tokyo 1951.

Neumann, Hanna: Varieties of groups. Berlin-Heidelberg-New York: Springer-Verlag 1967.

Neumann, J. von: (1) Zur allgemeinen Theorie des Maßes. Fund. Math. **13**, 73—116 (1929); — (2) On complete topological spaces. Trans. Amer. Math. Soc. **37**, 1—20 (1935); — (3) Proof of the quasi-ergodic hypothesis. Proc. Nat. Acad. Sci. U.S.A. **18**, 70—82 (1932); — (4) Rings of operators III. Ann. of Math. **41**, 94—161 (1940); — (5) Rings of operators. Reduction theory. Ann. of Math. **50**, 401—485 (1949).

Nikodym, O. M.: On transfinite iterations of the weak linear closure of convex sets in linear spaces. Parts A and B. Rend. Circ. Mat. Palermo **2**, 85—105 (1953); **3**, 5—75 (1954).

Nyokos, P., Schäffer, J. J.: Flat spaces of continuous functions. Studia Math. **42**, 221—229 (1972).

Orlicz, W.: (1) Beiträge zur Theorie der Orthogonalentwicklungen II. Studia Math. **1**, 241—255 (1929); — (2) Über unbedingte Konvergenz in Funktionenräumen. I, II. Studia Math. **4**, 33—37, 41—47 (1933).

Pelczynski, A.: (1) On the isomorphism of the spaces *m* and *M*. Bull Acad. Polon. Sci. **6**, 695—696 (1958); — (2) Linear extensions, linear averagings, and their application to linear topological classification of spaces of continuous functions. Dissertationes Math. (Rozprawy Mat.), No. **58** (1968); — (3) Any separable Banach space with the bounded approximation property is a complemented subspace of a Banach space with a basis. Studia Math. **40**, 265—289 (1971).

Peressini, A. L.: Ordered topological vector spaces. New York-Evanston-London: Harper and Row 1967.

Pettis, B. J.: (1) On continuity and openness of homomorphisms in topological groups. Ann. of Math. (2) **51**, 293—308 (1950); — (2) On integration in vector spaces. Trans. Amer. Math. Soc. **44**, 277—304 (1938); — (3) A proof that every uniformly convex space is reflexive. Duke Math. J. **5**, 249—253 (1939).

Phelps, R. R.: Lectures on Choquet's theorem. Princeton: van Nostrand 1965.

Phillips, R. S.: (1) On linear transformations. Trans. Amer. Math. Soc. **48**, 516—541 (1940); — (2) On weakly compact subsets of a Banach space. Amer. J. Math. **55**, 108—136 (1943); — (3) Integration in a convex linear topological space. Trans. Amer. Math. Soc. **47**, 114—145 (1940); — (4) Semi-groups of operators. Bull. Amer. Math. Soc. **61**, 16—33 (1955).

Pietsch, A.: Nukleare Lokalkonvexe Räume. Berlin: Akademie-Verlag 1965.

Porta, Horacio: Compactly determined locally convex topologies. Math. Ann. **196**, 91—100 (1972).

Price, G. B.: The theory of integration. Trans. Amer. Math. Soc. **47**, 1—50 (1940).

Pryce, J. D.: A device of R. J. Whitley's applied to pointwise compactness in spaces of continuous functions. Proc. London Math. Soc. (3) **23**, 532—546 (1971).

Ptak, V.: (1) On complete topological linear spaces. Čeh. Mat. Žur. (78), **3**, 301—364 (1953) (Russian with English summary); — (2) Compact subsets of convex topological linear spaces. Same Žur. (79), **4**, 51—74 (1954) (Russian with English summary); — (3) Weak compactness in convex linear topological spaces. Same Žur. (79), **4**, 175—186 (1954) (English with Russian summary); — (4) Two remarks on weak compactness. Same Žur. (80), **5**, 532—545 (1955) (English with Russian summary); — (5) On a theorem of W. F. Eberlein. Studia Math. **14**, 276—284 (1954) (English).

Rainwater, John.: Day's norm on $c_0(\Gamma)$. Proc. of Functional Analysis Week (Aarhus 1969), 46—50. Aarhus: Matem. Inst. Univ. 1969.

Retherford, J. R.: Some remarks on Schauder bases of subspaces. Rev. Roumaine Math. Pures Appl. **13**, 521—527 (1968).

Rickart, C. E.: Integration in a convex linear topological space. Trans. Amer. Math. Soc. **52**, 498—521 (1942).

Rieffel, M.: Dentable subsets of Banach spaces. In: Functional Analysis. Proc. of a Conference at Univ. of California, Irvine, 71—77. Washington, D.C.: Thompson Book Co. 1967.

Riesz, F.: (1) Some mean ergodic theorems. J. London Math. Soc. **13**, 274—278 (1938); — (2) Über lineare Funktionalgleichungen. Acta Math. **41**, 71—98 (1918).

Riesz, F., Sz.-Nagy, B.: Functional Analysis. New York: Ungar. 1955. Translated from Leçons d'Analyse Fonctionelle, 2nd edition. Budapest: Academia Kiado 1954.

Roberts, G. T.: The bounded-weak topology and completeness in vector spaces. Proc. Cambridge Philos. Soc. **49**, 183—189 (1953).

Robertson, A. P., Robertson, W. J.: Topological vector spaces. London: Cambridge University Press 1964.

Rockafellar, R. T.: Convex Analysis. Princeton, N. J. 1970.

Rosenthal, H. P.: On injective Banach spaces and the spaces $L^\infty(\mu)$ for finite measures μ. Acta Math. **124**, 205—248 (1970).

Rubel, L. A., Ryff, J. V.: The bounded weak-star topology and the bounded analytic functions. J. Functional Analysis **5**, 167—183 (1970).

Rubel, L. A., Shields, A. L.: The space of bounded analytic functions on a region. Ann. Inst. Fourier (Grenoble) 16 fasc. **1**, 235—277 (1966).

Ruston, A. F.: (1) Direct products of Banach spaces and linear functional equations. Proc. London Math. Soc. (2), **53**, 109—124 (1951); — (2) II. Proc. London Math. Soc. (3), **1**, 327—384 (1951); — (3) A note on convexity of Banach spaces. Proc. Cambridge Philos. Soc. **45**, 157—159 (1949).

Ryll-Nardzewski, C.: On fixed points of semi-groups of endomorphisms of linear spaces. Proc. Fifth Berkeley Symp. on Math. Stat. and Prob., vol. 2, part I, 55—61. Berkeley-Los Angeles: Univ. of Calif. Press 1967.

Schaefer, H.H.: (1) Topological vector spaces. New York: Macmillan 1966; — (2) Weak convergence of measures. Math. Ann. **193**, 57—64 (1971).

Schäffer, J.J.: (1) On the geometry of spheres in *L*-spaces. Israel J. Math. **10**, 114—120 (1971). — (2) On the geometry of spheres in spaces of continuous functions. J. Analyse Math. (to appear); — (3) —, Sundaresan, K.: Reflexivity and the girth of spheres. Math. Ann. **184**, 169—171 (1970).

Schatten, R.: A theory of cross-spaces. Ann. of Math. Studies **26** (1950).

Schauder, J.: (1) Zur Theorie stetiger Abbildungen in Funktionalräumen. Math. Z. **26**, 47—65, 417—431 (1927); — (2) Der Fixpunktsatz in Funktionalräumen. Studia Math. **2**, 171—180 (1930); — (3) Über lineare vollstetige Funktionaloperationen. Studia Math. **2**, 183—196 (1930).

Schoenberg, I.J.: A remark on M.M. Day's characterization of inner-product spaces and a conjecture of L.M. Blumenthal. Proc. Amer. Math. Soc. **3**, 961—964 (1952).

Schwartz, Laurent: Theorie des distributions, I, II. Paris: Hermann 1950, 1951.

Segal, I.: Decompositions of operator algebras I, II. Mem. Amer. Math. Soc. **9** (1951).

Semadeni, Zbigniew: Banach spaces of continuous functions. Monografje Matematiczne, vol. 55, Warsaw 1971.

Šilov, G.E.: Vvedenie v teoriyu lineĭnyh prostranstv. (Introduction to the theory of linear spaces.) Moscow-Leningrad 1952.

Silverman, R.J.: (1) Invariant linear functions. Trans. Amer. Math. Soc. **81**, 411—424 (1956); — (2) Means on semigroups and the Hahn-Banach extension property. Trans. Amer. Math. Soc. **83**, 222—237 (1956).

Silverman, R.J., Yen, Ti: The Hahn-Banach theorem and the least upper bound property. Trans. Amer. Math. Soc. **90**, 523—526 (1959).

Singer, Ivan: Bases in Banach spaces. Berlin-Heidelberg-New York: Springer-Verlag 1970.

Šmul'yan, V.L.: (1) Sur les ensembles régulièrement fermés et faiblement compacts dans les espaces du type (B). C.R. (Doklady) Acad. Sci. URSS **18**, 405—407 (1938); — (2) Über lineare topologische Räume. Mat. Sb. (49), **7**, 425—448 (1940); — (3) On some geometrical properties of the unit sphere in a space of type (B). Mat. Sb. (48), **6**, 77—89 (1939) (Russian, English summary 90—94); — (4) On the principle of inclusion in the space of type (B). Mat. Sb. (47), **5**, 317—328 (1939) (Russian, English summary); — (5) Sur les topologies différentes dans l'espace de Banach. C.R. (Doklady) Acad. Sci. URSS **23**, 331—334 (1939); — (6) Sur la dérivabilité de la norme dans l'espace de Banach. C.R. (Doklady) Acad. Sci. URSS **27**, 643—648 (1940); — (7) Sur la structure de la sphère unitaire dans l'espace de Banach. Mat. Sb. (51), **9**, 545—561 (1941); — (8) Sur les ensembles faiblement compacts dans les espaces linéaires normés. Comm. Inst. Sci. Mat. Mec. Univ. Harkov. (4) **14**, 239—242 (1937); — (9) On some problems in functional analysis. C.R. (Doklady) Acad. Sci. URSS **38**, 157—159 (1943).

Sobczyk, A.: (1) Projection of the space *m* on its subspace c_0. Bull. Amer. Math. Soc. **47**, 938—947 (1941); — (2) On the extension of linear transformations. Trans. Amer. Math. Soc. **55**, 153—169 (1944); — (3) Extension properties of Banach spaces. Bull. Amer. Math. Soc. **68**, 217—224 (1962).

Steinhaus, H.: Additive und stetige Funktionaloperationen. Math. Z. **5**, 186—221 (1918).

Stone, M.H.: (1) Applications of the theory of Boolean rings to general topology. Trans. Amer. Math. Soc. **41**, 375—481 (1937); — (2) The generalized Weierstrass approximation theorem. Math. Mag. **21**, 167—184, 237—254 (1948).

Straszewicz, S.: Über exponierte Punkte abgeschlossener Punktmengen. Fund. Math. **24**, 139—143 (1935).

Suhomlinov, G.A.: On extension of linear functionals in complex or quaternion linear spaces. Mat. Sb. **3**, 353—358 (1938) (Russian; German abstract).

Taylor, A.E.: The weak topologies in Banach spaces. Proc. Nat. Acad. Sci. U.S.A. **25**, 438—440 (1939).

Trèves, François: Locally convex spaces and linear partial differential equations. Berlin-Heidelberg-New York: Springer-Verlag 1967.

Troyanski, S.: (1) Equivalent norm in non-separable Banach spaces with unconditional bases. Teor. Funkciĭ Funkcional. Anal. i Priložen **6**, 59—68 (1968); — (2) On locally uniformly convex and differentiable norms in certain non-separable Banach spaces. Studia Math. **37**, 173—180 (1970—71); — (3) Example of a smooth space whose conjugate has not strictly convex norm. Studia Math. **35**, 305—309 (1970).

Tukey, J.W.: Some notes on the separation of convex sets. Portugal. Math. **3**, 95—102 (1942).

Tyhonov, A.: (1) Über die Erweiterung von Räumen. Math. Ann. **102**, 544—561 (1930); — (2) Ein Fixpunktsatz. Math. Ann. **111**, 767—776 (1935).

Uhl, J.J., Jr., Moedomo, S.: Radon-Nikodym theorems for the Bochner and Pettis integrals. Pacific J. Math. **38**, 531—536 (1971).

Vallée-Poussin, C. de la: Cours d'analyse infinitesimal. 8th edition. Louvain 1938 or New York 1946.

Veech, W.A.: Short proof of Sobczyk's theorem. Proc. Amer. Math. Soc. **28**, 627—628 (1971).

Wehausen, J.V.: Transformations in linear topological spaces. Duke Math. J. **4**, 157—168 (1938).

Wheeler, R.F.: (1) The equicontinuous weak* topology and semireflexivity. Dissertation. Columbia, Mo.: University of Missouri 1970; — (2) The equicontinuous weak* topology and semireflexivity. Studia Math. **41**, 243—256 (1972).

Whitley, R.J.: (1) An elementary proof of the Eberlein-Šmulian theorem. Math. Ann. **172**, 116—118 (1967); — (2) The size of the unit sphere. Canad. J. Math. **20**, 450—455 (1968).

Williamson, J.H.: Survey article on harmonic analysis on semigroups. J. London Math. Soc. **42**, 1—41 (1967).

Wilson, W.A.: A relation between metric and Euclidean spaces. Amer. J. Math. **54**, 505—517 (1932).

Yosida, K.: (1) Mean ergodic theorems in Banach spaces. Proceedings of the Imperial Academy, Tokyo **14**, 292—294 (1938); — (2) Quasi-completely continuous linear functional operations. Japan. J. Math. **15**, 297—301 (1939); — (3) Functional Analysis, 2nd Edition. Berlin-Heidelberg-New York: Springer-Verlag 1968.

Yosida, K., Kakutani, S.: Operator-theoretical treatment of Markoff's processes and mean ergodic theorem. Ann. of Math. (2) **42**, 188—228 (1941).

Zisler, V.: On some rotundity and smoothness properties of Banach spaces. Dissertationes Math. (Rozprawy Mat.) No. 87. Warsaw: Inst. Mat. Polsk. Akad. Nauk. 1971.

Index of Citations

Index of Symbols

Constants from Set Theory and Logic

$\emptyset$	empty set
$\cup, \bigcup$	union of sets
$\cap, \bigcap$	intersection of sets
$\setminus$	difference of sets
$\{x: P(x)\}$	the set of all x such that $P(x)$ is true
$\{x\}$	singleton x, the set whose only element is x
$\sum$	sum
$\times$	direct or Cartesian product of two sets or directed systems
$\prod$	direct product
$\in$	is an element of
$\Rightarrow$	implies
ω	directed system of positive integers
$\circ$	composition of functions

Non-alphabetical Symbols for Mathematical Operations

(on elements x, x', sets E, E', linear spaces L, L', linear functions T, T', or linear functionals f, f', or property of spaces (P))

Symbols which are Names or Abbreviations of Names

Symbols used as Labels for Properties or Conditions

In these abbreviations
w means weak(ly)
w^* means weak(ly)*,
ω means sequential(ly).

Subject Index

Ergebnisse der Mathematik und ihrer Grenzgebiete

52. Wilf: Finite Sections of Some Classical Inequalities. DM 28,—
53. Ramis: Sous-ensembles analytiques d'une variété banachique complexe. DM 36,—
54. Busemann: Recent Synthetic Differential Geometry. DM 32,—
55. Walter: Differential and Integral Inequalities. DM 74,—
56. Monna: Analyse non-archimédienne. DM 38,—
57. Alfsen: Compact Convex Sets and Boundary Integrals. DM 46,—
58. Greco/Salmon: Topics in $\mathfrak{m}$-Adic Topologies. DM 24,—
59. López de Medrano: Involutions on Manifolds. DM 36,—
60. Sakai: C*-Algebras and W*-Algebras. DM 68,—
61. Zariski: Algebraic Surfaces. DM 54,—
62. Robinson: Finiteness Conditions and Generalized Soluble Groups, Part 1. DM 48,—
63. Robinson: Finiteness Conditions and Generalized Soluble Groups, Part 2. DM 64,—
64. Hakim: Topos annelés et schémas relatifs. DM 48,—
65. Browder: Surgery on Simply-Connected Manifolds. DM 42,—
66. Pietsch: Nuclear Locally Convex Spaces. DM 48,—
67. Dellacherie: Capacités et processus stochastiques. DM 44,—
68. Raghunathan: Discrete Subgroups of Lie Groups. DM 56,—
69. Rourke/Sanderson: Introduction to Piecewise-Linear Topology. DM 42,—
70. Kobayashi: Transformation Groups in Differential Geometry. DM 52,—
71. Tougeron: Idéaux de fonctions différentiables. DM 69,—
72. Gihman/Skorohod: Stochastic Differential Equations. DM 88,—
73. Milnor/Husemoller: Symmetric Bilinear Forms. DM 42,—
74. Fossum: The Divisor Class Group of a Krull Domain. DM 44,—
75. Springer: Jordan Algebras and Algebraic Groups. DM 48,—
76. Wehrfritz: Infinite Linear Groups. DM 59,—

Prices are subject to change without notice